# Electronic fun with cheap eBay modules

**Spend nearly nothing to build your own amplifiers, power supplies, timers and flashers.**

# The author

Julian Edgar started his working life freelancing for photography magazines.

He then worked as a secondary school teacher for eight years, teaching senior humanities, before leaving teaching and becoming a full-time automotive writer.

He edited a national Australian automotive print magazine before becoming editor of *AutoSpeed*, an online car magazine. Along the way he wrote extensively for *Silicon Chip*, an electronics hobbyist magazine, while contributing articles to technical and automotive publications in Australia, the UK and the US.

This book is in part based on the author's very successful 'Electronic Building Blocks' column in *Everyday Practical Electronics* magazine.

Julian lives in a hamlet 80 kilometres north of Canberra, Australia. He spends much of the week playing in his home workshop – for the rest of the time, he works for a company providing training in high-level writing skills.

Other books by the author:

Automotive

> *21ˢᵗ Century Performance*, Clockwork Media, 2000
>
> *High Performance Electronics for Cars*, Silicon Chip Publications, 2004 (co-authored with John Clarke)
>
> *Amateur Car Aerodynamics Sourcebook,* CreateSpace, 2013
>
> *DIY Car Electronic Modification Sourcebook,* CreateSpace, 2013
>
> *DIY Testing of Car Modifications,* CreateSpace, 2013
>
> *Tuning Programmable Engine Management, CreateSpace, 2014*
>
> *Hybrid and Electric Cars Amateurs Sourcebook, CreateSpace, 2014*
>
> *DIY Suspension Development, CreateSpace, 2015*
>
> *Making custom curved trim panels for cars, CreateSpace, 2016*

Technical

> *Home Workshop Sourcebook, 2ⁿᵈ edition,* CreateSpace, 2016
>
> *Inventors and Amateur Engineers Sourcebook*, CreateSpace, 2013
>
> *Using the Brilliant eLabtronics Modules, CreateSpace, 2015*
>
> *DIY Loudspeaker Building, CreateSpace, 2016*

Business and Government

> *Writing Effective Arguments, CreateSpace, 2015*

Travel

> *Travels of a Wandering Amateur Engineer, CreateSpace, 2016*

# Contents

# A good quality 2.1 amplifier

**Run two speakers and a subwoofer with this incredibly cheap amplifier. Add a tone control board and you have a full-featured amplifier!**

Here's a really good 2.1 amplifier module that's at home in a lot of different situations.

Do you want a test amplifier for checking speakers? This one will do it.

Or maybe you want an amplifier to run a sound system in your child's room or home office?

You won't find anything better for the money. Or do you want something more serious? Then team the amp module with a tone control module!

*This three channel amplifier module is superb value for money. It has rated outputs of 50W x 2 plus 100W x 1. That's at 10 per cent distortion – but at half power, distortion is only 0.1 per cent!*

What we're talking about here is a class D amplifier module based on two TPA3116 chips. It has two output channels each nominally rated at 50 watts, and a third channel nominally rated at 100 watts.

You can use it as a stereo amplifier, or as a 2.1 amplifier – that is, driving stereo speakers plus the third channel driving a subwoofer.

To find the module, consult eBay using the search term "New 50W x2+100W TPA3116D2 2.1 HIFI Digital Subwoofer Amplifier Verst Board" or similar. It costs under $20, postage included.

In use, it's best to de-rate the module's maximum output power – at a full power of 50W per channel,

the TPA3116 spec sheet shows that THD+N is a high 10 per cent! However, at half power of 25W per channel, THD+N is 100 times better at only 0.1 per cent.

The module needs a 21V DC supply to provide full output power; in general use it's happy with anything from 12 – 25V. You can use any good plug-pack capable of suppling around 21V – a discarded (or new) laptop power supply is ideal.

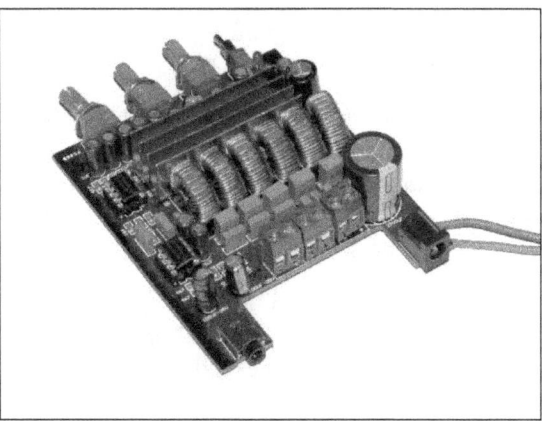

*The 1/8-inch stereo plug input is at left and the DC power socket is at the right. The red and green wires are ones I soldered directly to the board to provide power without needing the right plug.*

The module has a 1/8th inch stereo input socket, a DC power supply socket and a terminal block for the speakers. (Some versions have solder pads for the speaker connections.)

An on/off switch and three potentiometers are mounted on the front edge of the board. These knobs are: stereo speaker volume control, subwoofer volume control and master volume control.

I have been using this amplifier module now for about 6 months. It's outstanding – plenty of grunt, an excellent 102dB signal to noise ratio, able to drive 4 ohm loads, and protected against over-current and over-temperature.

And, talking about temperatures, the Class D architecture means that the tiny heat-sink shown in the pics is more than adequate – on normal music material, even playing loudly, it barely gets warm.

So how do you use the module?

One approach is to team it with second-hand speakers – for example, a discarded home theatre system comprising small satellite speakers and a subwoofer. Connect two satellite speakers to the stereo outputs, and the subwoofer to the third output. Using the two on-board pots, set the relative levels of the stereo pair and the subwoofer. Then, when these are set correctly, use just the main volume control to set the required listening level. (Or, if listening to MP3 files from a phone or similar, leave the main volume control at maximum and control the listening level with the volume control on the phone.)

The result is amazingly good – you'd think the amplifier cost $60, not around $20!

Another approach is to use the amp with just two higher quality speakers. Believe it or not, if you're not after audiophile quality sound, this tiny amp has performance that's good enough for the vast majority of listeners.

But what if you want a loudness button? Or separate bass and treble controls? In that case, welcome to the second of the modules.

*If you want to add tone controls to the amplifier, this module will do the trick. It has bass, treble, balance and volume pots (knobs supplied!), and an on-board loudness button.*

The second module is a tone control board. It has volume, balance, bass and treble pots. Furthermore, on the board itself is a loudness button. (This boosts bass and treble at lower listening levels.) On eBay, search under "LM1036 + NE5532 Stereo Preamp Preamplifier Treble/Bass Tone Board DIY Amplifier" to find it. It costs about $16, post included.

The output is via a 1/8th inch stereo socket (so use a cable with 1/8 inch stereo plugs at each end to connect to the amplifier module) and the input is via two RCA sockets. Power (12V) is supplied via either a DC-type socket or screw connectors (positive on the screw connectors is closest to the end of the board).

To supply this power, you could add a 12V plugpack or use a small power supply operating off the amplifier power supply – only a few milliamps is required for the tone control board.

As its name suggests, this board is based on the LM1036 tone control IC. The IC has low distortion (typically about 0.06 per cent) but its signal to noise ratio at 80dB is not as good as the amp module. If the Chinese-built module has followed the data sheet for the IC, bass (a centre frequency of 40Hz) and treble (16kHz) are variable by plus/minus 15dB.

The loudness button has a similar bass and treble boost, i.e. up to 15dB at low levels. However, this boost drops away nicely as the volume control is advanced. This module has one unfortunate glitch – if the loudness button is operated when the music is playing, a pop can occur in the speakers.

*The input is via a 1/8-inch stereo socket, and power is supplied via either the screw terminals or the DC socket.*

When I originally bought the amp module, I didn't dream that it would be of sufficient quality that I'd be using it during development of hi-fi loudspeaker systems – swapping only to a more expensive amp after my initial testing. Now the module has been retired from that function, and working with the tone control board, is powering a stereo pair of speakers in my home office. These two modules provide sound that literally only a few years ago would have cost you double or even triple...

# Phone blaster

**Build a powerful, quality sound amplified speaker for phone music.**

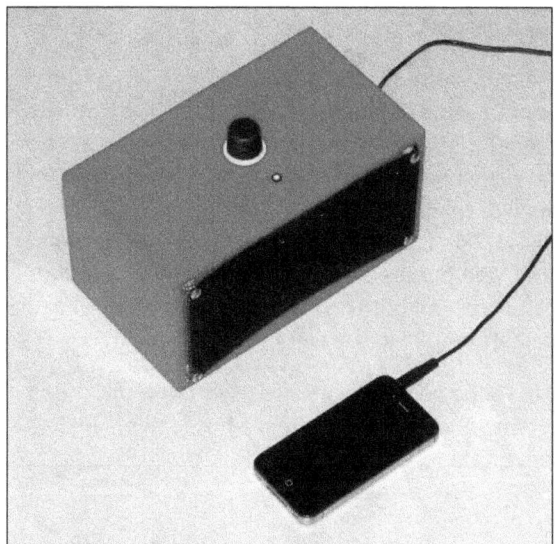

*This stereo, amplified speaker system has excellent sound and yet is cheap and easy to make. It's ideal working with a phone, as here. Seen on the top panel are the rotary on/off switch and a 'power on' LED.*

This project came from a requirement that nothing off the shelf could quite match. We live in country Australia, only about 50 miles from Canberra – Australia's capital – but still in the land of kangaroos, lots of sheep farms, and a school bus that takes well over an hour to get my 11-year-old son, Alexander, to the school at the local biggest town.  To while away the time, he reads books, reads (and plays games) on his tablet, watches the kangaroos and the kookaburras and the rabbits – and also, at times, wants to play music from his phone.

And Adam, the bus driver, is fine with some music.

But how to amplify the music from the phone, in a device that can also be carried in a school bag all day? Sure, there's a myriad of amplified speakers around, but is there a set that is compact, sounds reasonably good – and can satisfy the 'bus cred' that the statement "My Dad made this!" satisfies?

Not really, so I set to work.

The starting point was some marine-quality plywood, 7mm in thickness. This material formed the walls of the speaker enclosure. The front and back panels (also in 7mm marine ply) are attached to 18 x 18mm timber cleats that run around the inside of the enclosure. The outside dimensions of the enclosure are 210 x 110 x 120mm (width x height x depth).

I mitred all of the corners - and I wish I hadn't bothered going to the extra trouble as butt joints would have perfectly adequate. Marine ply (as opposed to normal plywood) was used to give as much stiffness as possible to the enclosure, while keeping it light.

The enclosure was sized so as to just fit in a stereo pair of 3 inch speakers – the enclosure has an internal volume of about 1.2 litres. The speakers were salvaged from a discarded home-style i-Phone amplified speaker system. They have large magnets and flexible roll surrounds. I could have developed a custom ported enclosure for these drivers, but measuring the Thiele-Small parameters of unknown speakers (while quite possible) is a lot of work that can be avoided if you're prepared to have some trade-off in ultimate low frequency response by using a sealed enclosure.

*Two 3-inch speaker salvaged from a defective home iPhone sound system are used, along with a cheap but very effective pre-built amplifier module. The system is shown here powered by a single 9V battery; if longer battery life is required, larger batteries can be used.*

The amplifier is available through eBay for around $8 – search under 'Tripath TA2024 amplifier module'.  In this application, its benefits are that is cheap, can work down to 7V, is efficient and fairly powerful. The amplifier is mounted on the inside of the rear panel, along with a 9V battery.

*The amplifier module has a 9-watt output and is excellent in this application.*

The speakers are mounted through the front panel, with the panel itself glued and nailed into position. The rear panel is removable, mounting on its inside the amplifier module and the 9V battery. Through a side panel is mounted a rotary on/off switch (rotary, so it's less likely to be bumped inside a school bag) and a blue LED used as a 'power on' indicator. (I used a LED pre-wired for 12V.)

*The inside of the sealed enclosure is packed with quilt wadding – very effective at improving the sound.*

Inside the enclosure is placed a generous amount of acrylic quilt wadding, serving the dual purpose of preventing acoustic reflections through the cones, and acting to increase the effective volume of the enclosure. Across the front is placed a stiff metal grille, protecting the fragile cones from fingers and odd missiles located in school bags (and school buses).

The audio signal is fed to the board via a cut-off 1/8th inch stereo adaptor cord (the cheapest way of getting a pre-wired plug), with the cord protected against pull-out by a knot.

Some wood primer and red enamel finished off the box with the required flair.

And how does it sound?

To be honest, I think it sound fabulous – far better than I'd expected. Some of that is just luck – the small drivers turning out to be good ones – but the stiff enclosure and decent amplifier module also play a big part.

When the system's frequency response is measured (using a free frequency generator app for the iPhone) the response is audible to just below 100Hz and Alexander can hear the output at 15kHz. (I'm too old to have a good high frequency hearing ability!) Admittedly, there are a few minor resonant peaks and humps along the way, but it still sounds better than all but the most high-end of small and portable amplified speakers.

So it's good enough to be also used as a picnic sound system, let alone to entertain a bus-load of kids!

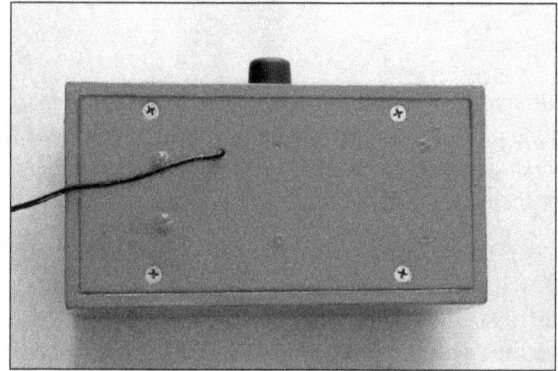

*The rear panel shows the input cable and four screws that allow the back to be removed for battery replacement.*

**Battery life?**

So, with a 9V battery, how long will the battery last? Well, that depends on how loudly the system is played. I measured an 80mA current draw at a 'moderate' listening volume. That would make the 9V alkaline battery last something like 6 hours or so. Used 20 or 30 minutes a day, that should give reasonable battery life.

However, using a rechargeable battery (e.g. an 11.1 lithium-polymer pack and associated charger) would be a better bet if the system is to be played a lot.

# General purpose 12V timer

## A simple timer with a relay output and hundreds of uses.

*Need to operate something for a period at the press of a button? It doesn't get much cheaper or easier than this! Available now on eBay for just $14 (including delivery to your letterbox!) is this general purpose 12V timer. Search under "Adjustable Delay Timer Module". The module is adjustable for periods from 1 second to 3 minutes, comes with an on-board relay.*

*Measuring only 56mm x 26 x 23mm (L x W x H), the tiny module has a LED, relay and multi-turn adjustable pot. There are two terminal strips, one at each end. At one end there are the connections for:*

- *12V and ground*
- *Trigger input*

*At the other end there are the relay connections for:*

- Common (x2)
- Normally open
- Normally closed

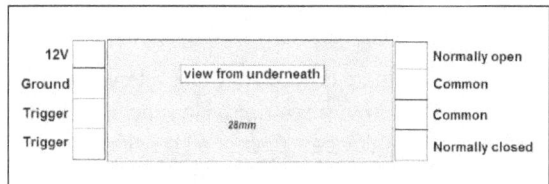

*Viewed from underneath, the connections look like this. Note that when being viewed in the correct orientation, the '28mm' written on the bottom should be the 'right way up'.*

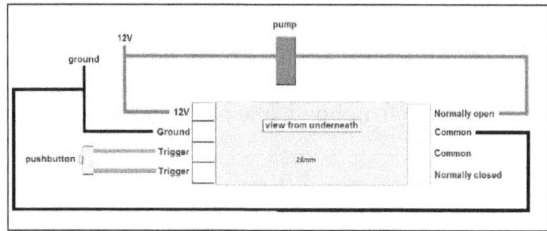

*The connections are easy. Let's say that you want to run a 12V powered device (e.g. a pump) for 30 seconds with just one push of the button. Connect:*

- *12V*
- *Ground*
- *The two trigger wires to a normally open pushbutton (doesn't matter which way around these wires go)*
- *One side of the pump to 12V*
- *The other side of the pump to the normally open relay terminal*
- *The common of the relay terminal to ground*

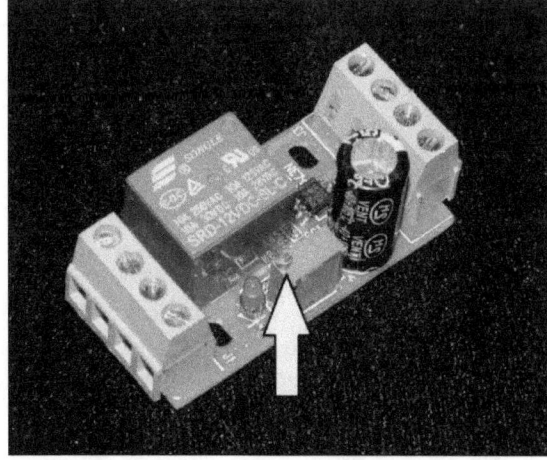

*The multi-turn adjustable pot (arrowed) changes the timed period. As viewed from above, turn the pot clockwise to make the timed period shorter.*

# Electronic stethoscope for machinery

## Diagnose mechanical noises easily.

Here is a really neat device that will allow you to listen to all sorts of strange noises in machines and other mechanical devices. Best of all it will cost you less than $8 for the main part of the device – just add some headphones and shielded microphone cable.

The Electronic Stethoscope is based around the 'Listen Up' portable sound amplifier available from many eBay suppliers. The device is designed for people who are hard of hearing but don't want to wear a hearing aid. It comprises a small amplifier with a built-in microphone, and pair of earphones.

As supplied, the device is not particularly high quality and doesn't even work very well in its intended function. But remote-mount the microphone and plug some decent headphones into the jack and it works extremely well!

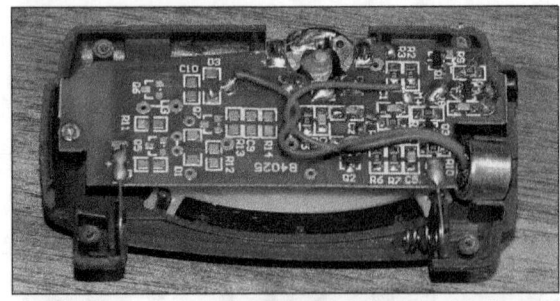

The first step is to buy 'Listen Up', some cable (I used shielded two-core microphone cable) and a 10-amp battery clip (you can also cut a clip off any old battery charger – that's what I did). Open the box by first taking off the end caps (one covers the battery – a single AAA cell) to reveal four small screws. With the box open, it should look like this.

Note the microphone (arrowed) – both the fact that it is a large and relatively good quality unit, and that it is attached to the printed circuit board with flying leads. Unsolder the microphone leads and solder to the board the conductors in the new shielded microphone cable.

Here are the new wires soldered to the original locations. I also soldered the braid of the cable to the negative connection of the battery (arrowed), to provide better shielding of the signal. Make a suitable hole for the cable to escape and then close up the box.

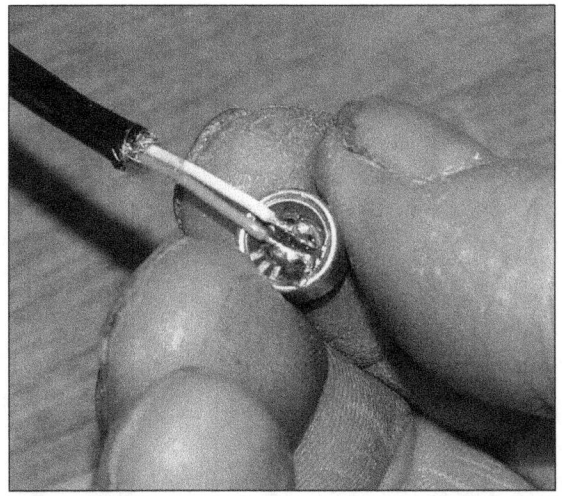

Solder the conductors in the other end of the microphone cable to the microphone. Keep the polarity the same as original. I chose to use a 4 metre length of cable as I wished to be able to monitor car engine noises from inside the cabin.

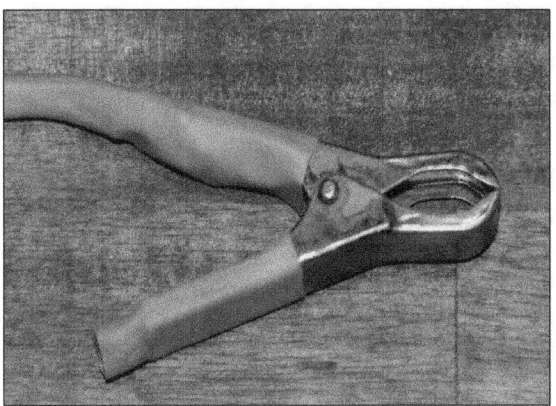

Use hot melt glue or similar to mount the microphone to the inside arm of a metal battery clip, then cover in heatshrink. If you desire, you could instead glue the microphone to the end of a steel rod – placing the other end of the rod against the machinery will work well if you want to accurately pinpoint where a particular noise is coming from.

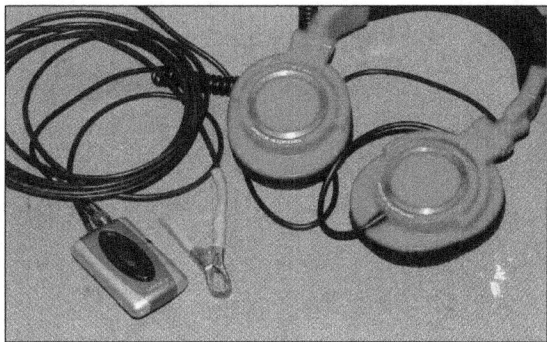

As mentioned, the earphones supplied with the 'Listen Up' device are pretty poor, so supply your own. High quality, fully enclosed headphones will work best – but any decent quality earphones should also be fine. With fully enclosed headphones, the sound quality is excellent.

### Using It

Using the electronic stethoscope is very simple. You simply clip to the microphone to whatever you are interested in listening to. Noises are transmitted through the metalwork directly to the clip and microphone, making the device extremely sensitive.

### Other fun uses

There are other uses for the 'Listen Up' device modified as described here.

My son, age 11, was entranced by the fact that he could use a long cable for the microphone and eavesdrop on his parents. (He also talked about taking it to school and bugging the staff room. Wonder where he gets these ideas from?!)

Given that the project involves some simple disassembly and soldering and so is electronically quite educational, I bought him one of the devices for his birthday.

# One very smart LED

## Battery monitoring like you've never seen it before.

I have to say that I am impressed - here is a really smart gadget that does exactly what its UK manufacturer says it will do.

*At a glance, the battery monitor just appears to be a 10mm white LED. However, its functionality is much greater than it first appears.*

So what is it then? Well, at first appearance it looks just like a 10mm LED mounted in a bezel. But then when you look closer, you'll see a tiny pushbutton – and if you pull the assembly from the bezel, you'll also see a programmable chip and a few other components.

In fact, what we have here is a 6, 12 and 24V battery monitoring LED that is user-programmable to run one of six different in-built maps. The single LED can show green, red, yellow (a yellow that actually looks more orange) and white (off). The LED can also flash at different rates. The voltage is monitored as a rolling average over 2 seconds, is claimed to be accurate to 1 per cent, and will operate over the range of 3.8 – 30V.

Wiring is as simple as you can get – red to positive and black to negative. That's it. Here we'll describe the monitor as working with a 12V battery, for example in a car.

### The maps

So what are the different in-built battery monitoring maps that are available? There is one for pretty well every use you can think of.

Here are the different maps:

### Map 1: Battery Voltage Monitor

This is the factory default voltage indicator mode. Low distraction, minimum of colour changes in normal operation, suitable for vehicle use, such as motorcycles, cars, boats, campers, etc. By flashing and using different colours, it shows eight different voltage ranges from 10.5 to 15+ volts.

### Map 2:  Vehicle Charge Indicator

This map illuminates the LED green when under charging conditions, i.e. when the vehicle alternator is working. Yellow and red will show if the battery is discharging. This mode monitors seven different voltage ranges from 10.5 to 15+ volts.

### Map 3: Vehicle Monitor, includes fake alarm

This is great for motorcycles, and vehicles stored long term. When riding/driving and charging, the LED is steady green. 30 seconds after charging stops (i.e. the vehicle is parked), the unit will enter low current mode to show battery status while the vehicle is in storage. The LED will blink green, yellow or red to show stored state battery condition. An added benefit is that LED blinking looks like a vehicle alarm. This mode has only a very small current draw (0.5mA) from battery.

### Map 4: Charge Indicator (Stealth Mode)

This is similar to mode (2) but the LED is not illuminated under normal charging conditions. That is, the LED is blank in normal operation. Yellow and red illuminations signal charging faults or discharging battery.

**Map 5: High Res 10 step voltage monitor**

This mode is a high resolution mode where maximum resolution is important and colour changes and flashing are not distracting. This mode monitors 10 different voltage ranges from 10.5 to 15+ volts.

**Map 6: Minimal Monitor**

This mode uses a simple low current (less than 0.5mA), three colour battery status monitoring. A short flash every 2 seconds indicates current state, with 5 different voltage ranges from 10.5 to 15+ volts.

**Different voltages**

The tricks of the device don't end with the different maps. As mentioned, you can also configure the LED to work on battery voltages of 6, 12 or 24 volts!

**Using it**

So is it all good news? Well, we initially found the instructions rather hard to understand - especially in the area of mode set-up.

Let's take a look at this aspect.

A delivered, the LED is set to Mode 1. To move to the next mode (i.e. in this case Mode 2) you do the following:

1. Power-up the LED

2. Hold in the pushbutton

3. Wait until the LED flashes green

4. Release the button

Now here is the trickier bit. To confirm what mode you have now set, turn off power and then re-connect it.

In this case you would expect to see three green flashes (indicates the LED is still in its 12V battery monitoring setting), a pause, and then green flashes (indicates Map 2).

The maps settings (the second lot of flashes) are as follows:

Map 1 – red flashes

Map 2 – green flashes

Map 3 – yellow flashes

Map 4 – red and then green flashes

Map 5 – yellow and then red flashes

Map 6 – yellow and then red flashes

(...and remember we said that yellow looks more like orange!)

Once you've sorted this aspect out, changing the voltage to 24V or 6V monitoring is straightforward. The procedure is:

1. Power-up the LED

2. Hold in the pushbutton

3. Wait until the LED flashes green, then red

4. Release the button

The voltage mode will switch to the next, so 12V to 24V to 6V to 12V – and so on. Then examine the initial flashes after switch-on on the basis of the following:

6V – red flashes

12V – green flashes

24V – yellow flashes

**Uses**

This little unit is ideal for battery monitoring in cars (especially those driven irregularly), motorcycles, 6V and 12V portable battery powered equipment – you name it, if it has a 6V, 12V or 24V battery, this programmable monitor will probably suit. It is also ideal for camping and remote area power supply systems running off battery or solar power.

**Cost**

The unit costs $22 plus postage. It is made by Gammatronix in the UK and is available via eBay. Search for '6v, 12v, 24v Programmable LED Battery level voltage monitor meter indicator'.

# PIR sensor module

**Detect movement with this sensitive module that has an adjustable relay output.**

Here's a great off-the-shelf module that can be used in 'serious' applications like security or automatic light operation – or for fun applications like catching your son or daughter sneaking into your lolly cupboard!

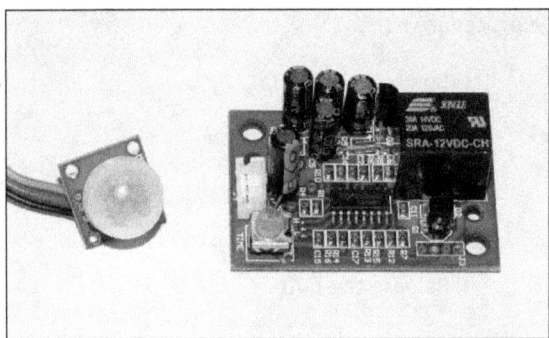

*The PIR sensor module comprises a small PCB, 45 x 30mm. It connects to a supplied sensor via a 150mm long cable. A relay output is provided so it is easy to use the module to trigger lights or buzzers.*

The module comprises a small board 45 x 30mm. The sensor, mounted on its own 15mm square board, comes supplied with about 150mm of cable and is fitted with a plug. A matching socket is provided on the main board, so connecting the two requires simply plugging the sensor into the socket. (Note that this socket is the one nearest to the adjustment pot.)

Another socket, located near to the relay, is provided on the board and this comes with a three-wire, flying lead cable. The wires are colour-coded:

- Red – DC +12V input

- Black – negative

- Yellow – DC +12V output

The output is provided by an on-board relay that's nominally rated at 20A. However, the board tracks would not cope at all well with this much current, so perhaps don't load it with more than an amp or so.

Also provided is an on-board pot. This does not, as you might first expect, adjust the sensor sensitivity but instead alters the time the output stays energised after the sensor has last detected movement. This output is variable in the range 15 seconds to 30 minutes. Note that clockwise = shorter duration.

The standby current is claimed to less than 50uA – and we measured 42uA when the module was powered from a 9V battery. That means that battery life when the device is in an un-triggered state should be excellent.

The sensor is sensitive and, if located in an upper corner, will detect movement anywhere in a large room.

We chose to power the module from a 9V battery and trigger a high intensity LED. That very simple set-up illuminates cupboards when they are opened. Replace the LED with a buzzer and the device can be used to detect people raiding lolly jars and the like!

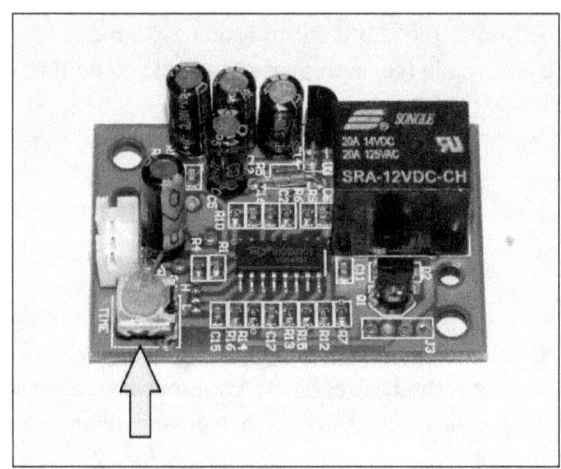

*This on-board pot allows the output time of the relay to be set. Minimum actuation time is 15 seconds and maximum is 30 minutes.*

A great module with plenty of real-world uses.

This module will cost you less than $8, delivered to your letter box. Search under "DC 12V PIR sensor High sensitivity PIR Module with Relay control".

# Pulsing timer module

**Here's a great electronic module that can be used to pulse an output. It is widely adjustable, with both the 'on' and 'off' times able to be set separately.**

*The module uses a relay output that is able to drive high current loads (up to 10 amps). It's also incredibly cheap – less than $16 delivered to your letterbox! (Do an eBay search under '12V DC Circulate Time Delay Relay module'.)*

*The fully constructed module is about 56 x 30mm. At one end of the board it has inputs for power and ground, and the other end has three relay connections – one for common, one for normally open and the other for normally closed. There are two pots mounted on the board – one controls the 'off' time and the other the 'on' time. A red LED glows whenever power is applied, and a green LED turns on when the relay is activated. There is also a configurable link on the board. Placing the link on the board feeds 12V to the 'common' terminal of the relay. This makes wiring much simpler in many applications, because the load can be connected between the other relay terminal and ground (more on this in a moment).*

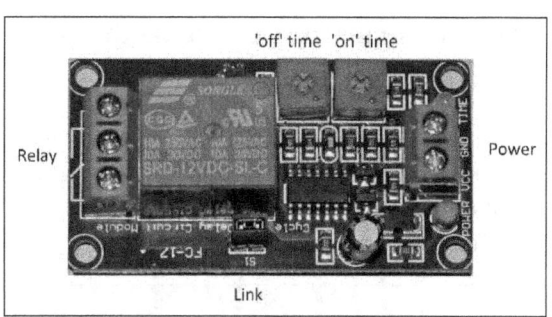

*The easiest way to see how the module works is to connect power and ground to the appropriate terminals. Rotate both pots fully anti-clockwise. Turn on power and the red LED will immediately light. With the pots set as described, the green LED will flash (and the relay) click, with the 'on' and 'off' times both being 1 second. This is the shortest output time available.*

*Rotate the 'off' pot a fraction clockwise. The output will still be activated for one second but this might now occur only every 5 seconds. And if for example you wanted the output to be activated at 5 second intervals for the longer period of 10 seconds, you'd turn the 'on' pot a little more clockwise.*

*You can see that both the frequency and duty cycle can be adjusted in this way. On the sample module, the 'off' time was adjustable from 1 – 120 seconds, and the 'on' time from 1 – 40 seconds.*

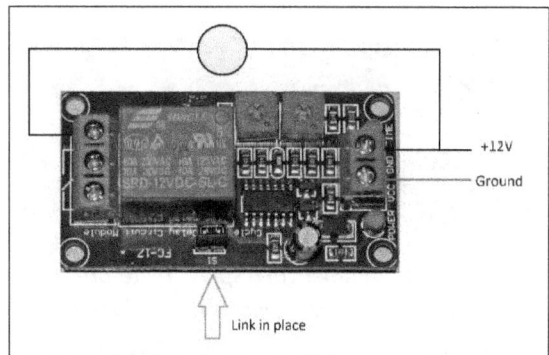

*This pic shows the easiest wiring connections, where a lamp is being flashed. Power and ground are connected as shown. The light is wired between the normally open terminal of the relay and ground. The link (arrowed) is in place that directs +12V to the common terminal of the relay. The light then flashes whenever power is applied.*

# Temperature controller and display

## A brilliant temperature controller that's so cheap it's unbelievable.

Here's an incredibly cheap temperature controller. How cheap? Try under $14 delivered to your letterbox. To find it, search on eBay under "DC 12V Digital Temperature Controller Thermostat C".

*The module is 78 x 71 x 29 mm (L x W x H) and uses a display window that requires a cut-out 70 x 28mm. It has a mass of 110 grams. It uses a LED display that shows temps up to 100 degrees C to one decimal place (e.g. 35.6), and above 100 degrees C in single units (e.g. 105). The update rate is fast (about 3 times a second) and the sensor is very responsive to changes in temperature.*

*In addition to the numerical display, there are two individual LEDs. One shows when the set-point has been exceeded. (The set-point is the temp at which you've set the device to activate its output.) This LED has two modes – steadily on when the relay is activated, and flashing when the set-point has been passed but the module is running an inbuilt delay before turning on the output. (You can vary this delay time – more on this in a moment.)*

*The other single LED shows that the display is indicating the set-point temperature.*

*On the face of the instrument are four push buttons – up/down arrows, Set and Reset.*

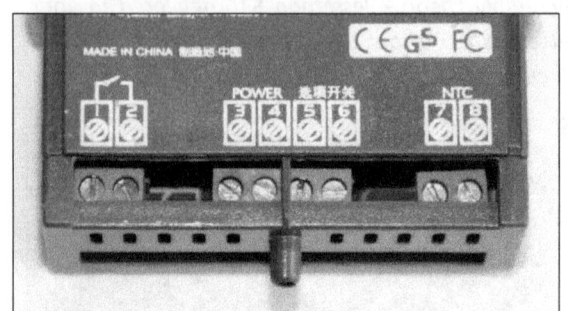

*Wiring connections are by means of screw terminals on the rear of the module.*

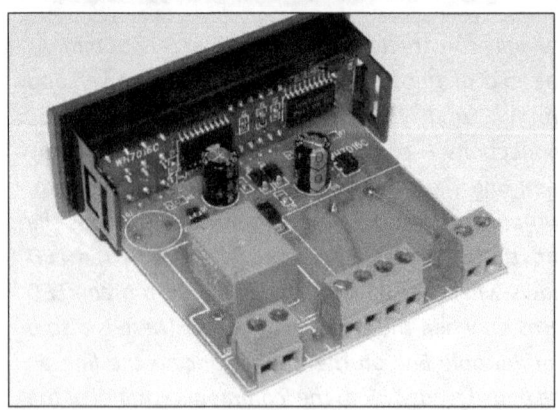

*The module doesn't look at all cheap and nasty – in appearance it could easily be an expensive instrument…. and that sentiment also applies to the internal build quality.*

## Making Connections

- **Displaying Temperature**

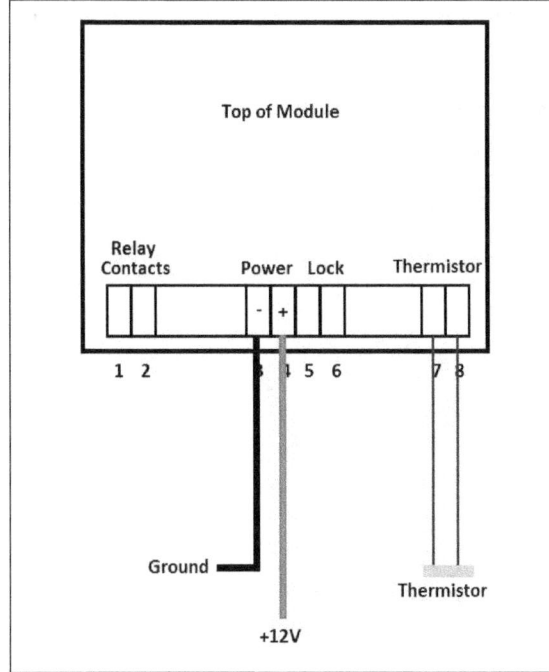

*Here is the wiring for a simple temperature display.*

The simplest use of the instrument is to display just temperature. This requires only four wiring connections and no menu configuration.

Power (12V nominal) connects to Pins 3 and 4 – ground to pin 3 and positive to pin 4.

The NTC (Negative Temperature Coefficient) sensor that is provided connects to pins 7 and 8 – it doesn't matter which wire goes to which terminal. With these connections made, the display should come alive and show the temperature at the sensor.

(Note: the thermistor wiring can be extended as required.)

The default values programmed into the instrument mean that straight out of the box it will work fine as a digital thermometer.

- **Controlling an output**

The module is fitted with a 5 amp relay. This means it can be connected directly to low voltage buzzers, fans and warning lights.

To get a feel for how the control system works, it's a good idea to play with it before installation. Let's take a look at how it can be set up.

Pressing the Set button briefly changes the display to show the set-point temperature. This setting can be altered by pressing the up and down keys. When done, press the Set key again or simply wait a few seconds and the display reverts to the current temperature.

Pressing the Set button for 3 seconds brings up a second menu. Different parameters can be selected by pressing the up/down keys. To change the selected parameter, press the Set key a second time then make the adjustments with the up/down keys. Whatever setting is selected is retained in memory, even if power is lost.

The available parameters are:

HC – this menu configures the module to either turn on its relay when the temperature exceeds the set-point ('C' mode), or turns on the relay when the temp falls below the set-point ('H') mode.

d – this sets the difference in temp between switch-on and switch off. (This is sometimes called the hysteresis.) By using the up/down keys, you can set this anywhere from 1 degree C to 15 degrees C. This is a very powerful control that can make a huge difference to how the system functions.

L5 – this is the minimum temperature the set-point can be configured. Normally, this would not need to be altered from its minus 50 degrees default.

H5 – this is the maximum temperature the set-point can be configured. Normally, this would not need to be altered from its 110 degrees default.

CA – this function allows you to correct the temperature display by adding or subtracting 1 degree units from the displayed reading.

P7 – this function is used when in C mode you don't want the output cycling on and off at short intervals. The setting can be anything from 0 – 10 minutes. It example, if it is set to 1 minute, after the relay has activated once, it will not activate again until a minute has passed – even if the temp set-point has been tripped. In most uses you would set this to zero.

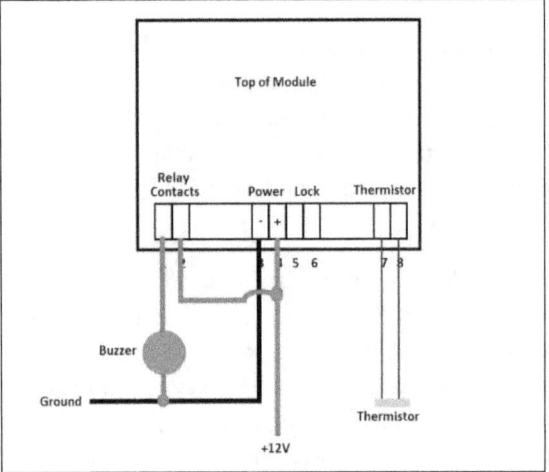

*Here is the wiring for an over-temp alarm buzzer.*

**Conclusion**

For the money, this is just a phenomenal instrument. Having an accurate and fast-response digital thermometer is just great – but being able to smartly control a relay output is just amazing.

The icing on the cake is the adjustable hysteresis and delay to prevent cycling.

Oh yes, and the fact you can calibrate it…

Gosh, what a buy!

---

# Must haves

*These digital voltmeters are cheap and with their two-wire design, easy to hook-up. They work over the range of about 3 – 30V DC and so are ideal for fitting to battery chargers and for monitoring the voltage of rechargeable batteries. Expect to pay only $2 each.*

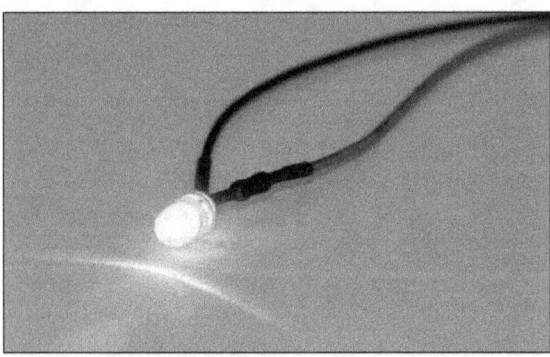

*I bought a whole bunch of these blue LEDs and have found them more useful than I expected. They're high intensity and come fitted with a dropping resistor allowing them to be powered straight from 12V. However, in use they're easily bright enough to be used on voltages down to 3V, so allowing them to be used on most low voltage equipment.*

# Ultra-low current LED flasher

**Incredibly low current draw and it will cost you almost nothing.**

Here's a project that ticks all the boxes – it costs nearly nothing, is very easy to build, and is extremely useful. So what is it? It's an ultra-low current LED flasher. So what use is that then?

*Any small battery-powered clock can be used as the basis of this project. You can buy them new very cheaply or salvage the workings from a discarded clock.*

Well, if you need to have a 'power on' indicator on battery-operated equipment, this LED will consume far less power than a conventional LED. That way, the 'on' indicator is contributing only fractionally to the flattening the battery.

Or take an indicator that needs to provide a security warning – like the flasher now fitted to many car dashboards or door cappings. Again, the requirement is for an indicator that takes very little current.

Where you want to be able to find a camping lantern in a dark tent – just fit this flasher and you'll always know where it is. And you can leave it flashing all night without taking any more than a tiny current draw from the battery.

So wherever warnings or indicators are needed, or the requirement is for a flashing LED that can operate in a low power environment, here is the LED flasher that you need.

And best of all, it will take you only a few minutes to build!

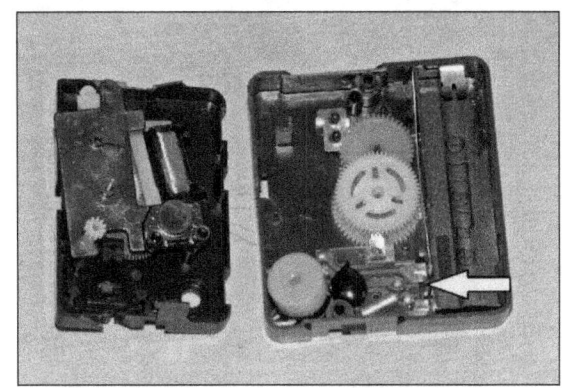

*Disassemble the clock until you can remove the circuit board (arrowed). Before you do so, take special note of the polarity of the battery connections and which solder pads connect to the solenoid coil.*

**The building block**

The circuit board for this project is taken straight from a battery-operated clock. You can buy one new on-line for just a few dollars (and that includes delivery) or purchase one from a discount store. Or you may well have an old battery-operated clock around the place that you can recycle.

Remove the circuit board from within the clock module. Carefully study (1) the polarity of the power connections to the board, and (2) the connections for the external solenoid coil that powers the clock mechanism. Many small alarm clocks also have a remote piezo buzzer – you don't need this so it can be removed by snipping its wires.

So how do you make this module into a flasher? Simply connect a high intensity LED to the solder pads that once went to the solenoid coil. Then connect a 3V source to the power connections, observing the original polarity.

It doesn't matter which way you connect the LED, and despite the clock originally being powered by 1.5V, it will work fine on 3V.

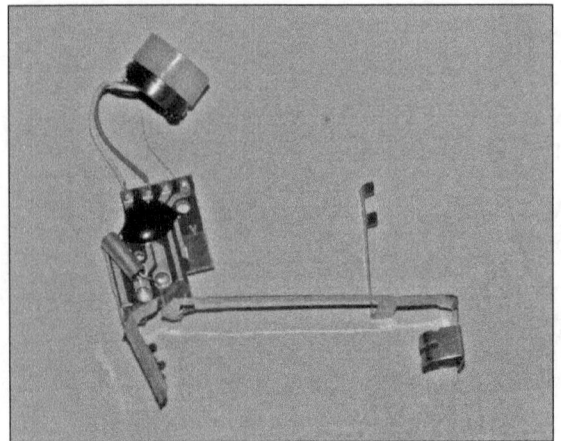

*Remove the electronics. In this case, the clock had an alarm – we don't want it in this application, so the buzzer wiring can be cut off. This PCB design uses battery clips that can be slid off the board – other designs will require the battery wiring to be unsoldered.*

But hold on?! It doesn't matter which way you connect the LED – how can that be so?

The clock module outputs a pulse every second – but one is negative-going and the other positive-going. The LED lights only when its connection polarity matches the direction of the pulse. Furthermore, because the pulse is so short and the internal current source appears limited, you can drive pretty well any high intensity LED directly from the output without using a current limiting resistor. (If you are worried by that, you can of course insert a series resistor.)

**Flashing**

The LED will flash once every two seconds for a 31 millisecond pulse – a duty cycle of just 1.56 per cent. (At a measured 20mA current, you can now see why battery life is so long!) If you want a faster flash rate, just parallel another LED in reversed polarity to the first. The LEDs will flash alternately, a LED lighting once per second. You can also use two different colour LEDs – for example, you can have a green/red flasher.

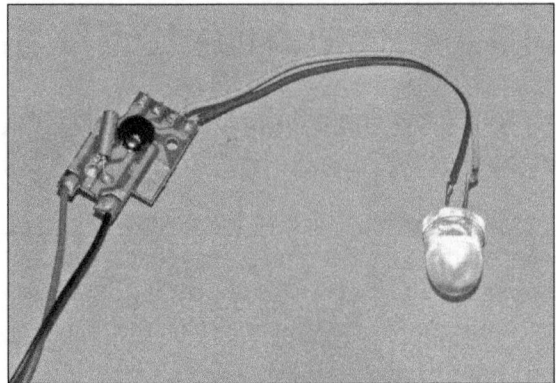

*Connect 3V of battery power to the original power terminals (use the correct polarity!) and solder a LED directly across the solenoid coil outputs. Use flexible thin insulated wire for these connections.*

Be very careful when soldering to the board. Always use small diameter flexible wires rather than solid-cored copper (as is used on the LED leads). If you don't use flexible wires, it's very easy to lift the solder pads off the board by bumping the connection.

The LED you use can be as large as the 10mm design used here, or as small as a 3mm unit. If you want a prominent warning, use a 10mm LED and cover it in a tube of semi-translucent plastic – I used the cap from a thick marker.

**How long will it flash?**

So how long will the LED flash for? Incredibly, it will keep flashing for literally years on good quality AA batteries, and for at least a year even on average quality batteries.

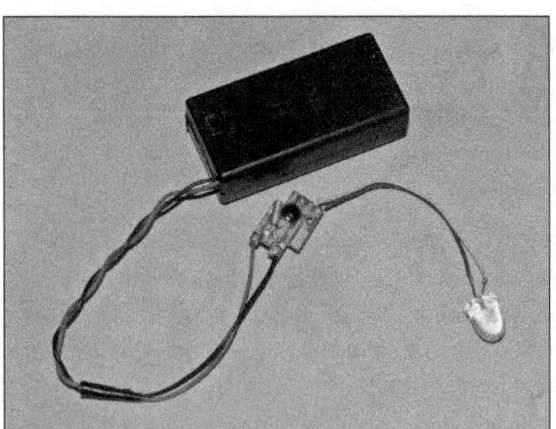

*The completed LED flasher, ready for mounting in a box.*

# Incredibly cheap voltage switch

**An amazingly cheap voltage switch that activates a relay when the input voltage reaches the required level.**

Available through eBay (do a search under "DC 12V Dual Wire Actuation Type Photoswitch Sensor Relay Module"), this module will cost you less than $10 delivered to your letter box.

As indicated in the above description, the device is sold as a photo-switch – a device that switches a relay on the basis of different light levels. The module comes with a light sensor that can be easily remote mounted. In fact, it's a perfect module for switching on lights when it gets dark.

But hold on, what's this got to do with a voltage switch?

The trick is this: if you unplug the light sensor and instead feed a variable voltage to one of the exposed pins, the module then becomes a universal voltage switch, suitable for monitoring voltages in the range of about 0.5 - 5 volts. The module runs on 12V.

**Testing**

We suggest that you first test the module as a light switch – just to make sure everything is working as it should.

*Here is the module as it comes from the supplier.*

*Remove the link (arrowed) and place it in a secure place – it's easy to lose it.*

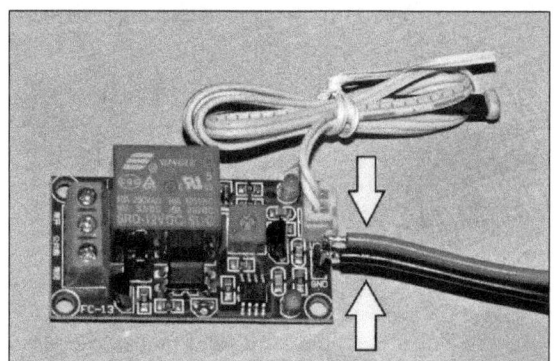

*Connect +12V and ground wires as shown here. Switch on power and the red LED should light, indicating that the module is powered-up.*

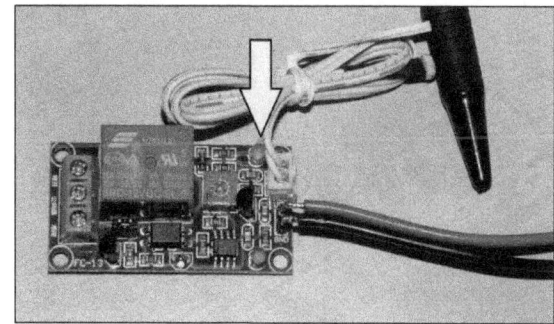

*Shade the light sensor and the green LED (arrowed) should light and the relay click. Adjust the pot and you will be able to adjust the light level at which the relay operates.*

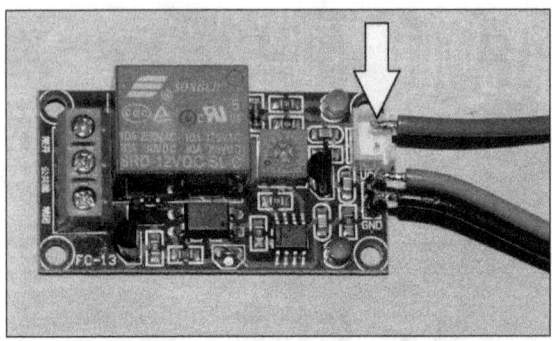

Now carefully unplug the light sensor. Cut away part of the connector housing for the light sensor so that you can solder directly to the pins. Solder your input signal wire to the pin furthest from the power connections, as shown here by the arrowed wire. This is the wire that connects to whatever voltage source you are sensing.

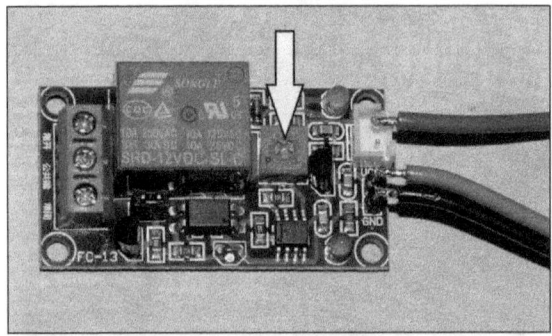

Rotating the pot anti-clockwise increases the voltage at which the module switches. When the pot is adjusted correctly for the application, a dab of nail varnish can be used to hold it in the chosen position.

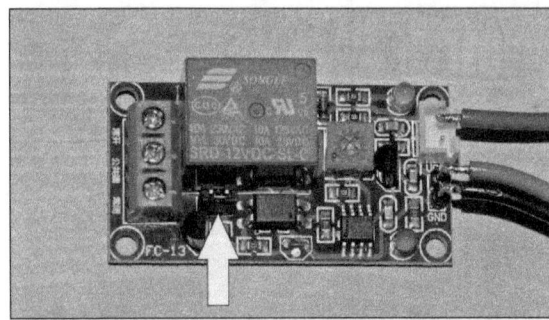

The relay terminals are configured (from, top to bottom): Normally Open, Common, Normally Closed. If the link is placed as arrowed (that's the link you put aside earlier), 12V is fed to the Common relay terminal. This makes wiring much simpler when you want to operate a low current load like a buzzer or lamp.

Here is how the module is configured to turn on a light when the monitored voltage **exceeds** the pre-set level.

Here is how the module is configured to turn on a light when the monitored voltage is **below** the pre-set level. (Note how the green LED is now off.)

And if you can afford a bit more, this different voltage switch is also available. It is digitally programmable and constantly displays the measured voltage on the LED display. Search under "Voltage Tester Monitor Charge / Discharge Over / Under Voltage Overload Protect".

# A new control system for a spot welder

**Using cheaply sourced parts to revolutionise the usefulness of an old workshop machine.**

I'd now like to cover something a little different. Rather than look at individual products purchased on eBay (or other online sites), I'd like to cover a complete project built from parts sourced very cheaply.

It's not a project that each of you will be rushing off to do for yourself, but it's interesting for two reasons. Firstly, it shows just what is possible with such a vast range of electronic modules and components now available, and secondly, it shows that projects that would have once not been viable because of cost have now not only become viable, but in fact cheap! Such is the change in the economics of our hobby over the last few years...

*This old spot welder has a high current capability and is built like a tank. However, as purchased, the welding duration was controlled only by a mechanical switch operated by a foot pedal. This made setting the welding duration a bit hit-and-miss.*

### The project

I have an extensive home workshop with a variety of metal- and wood-working tools. For example, I am lucky enough to own oxy-acetylene, MIG and TIG welders. However, I decided that additionally I wanted a spot welder, but found that these welders have not come down in price in the way that other welders have.

I therefore decided to buy an old, second-hand spot welder. The machine, that probably dates back to the 1950s, weighs a considerable 76kg when mounted on its stand. Cost was $700, about half the price of a new welder with similar secondary current.

The old welder was originally designed so that when the foot pedal was pushed down, the electrodes clamped the workpiece. Press further, and a switch was triggered that activated the current. Because the arms were closed by a foot pedal, the welding time was set only by movement of the operator's foot.

In spot welding, the duration of the weld time is very important. For example, when welding thin gauge metal, too short a weld time (e.g. 0.2 seconds) gives a weak weld. On the other hand, too long a weld time (e.g. 0.5 seconds) and the metal becomes overheated, with liquid metal expelled from the weld. I therefore decided to add an adjustable electronic timer that would allow the welding duration to be accurate and repeatable.

These welders draw a lot of current from the mains and develop secondary currents measured in thousands of amps. So what would happen if you went on a welding frenzy? The answer would be the welder's transformer would get hot. A fan to cool the transformer was then added.

Finally, with these simple transformer-type spot welders, the amount of secondary current (the current that does the welding) depends on the mains voltage, as received by the welder. And because of the very high current draw, the voltage drop can be considerable. Throw in a mains voltage where I live that can vary by 20 volts or more (and that's off-load!), and I decided to add a mains voltage display.

When looking for one of those, I realised that I could add a mains current display at little additional cost – so I bought a combined voltage / current meter.

• Micro-switch automatically triggers the timer when full electrode clamping pressure has been reached

• Over-ride pushbutton can be used by the operator to give a manual weld time

• New control panel has main on/off switch, 'welding occurring' pilot light, manual weld over-ride pushbutton, adjustable digital timer, and combined mains voltage and mains amperage meters.

• Small power supply generates 12V for solid state relay control, digital timer and pilot light.

Let's take the major parts one at a time.

### 1. Solid state relay

When set to its highest power, the welder can draw up to 28 amps, and it takes a 38-amp gulp when first starting a weld. These are very high currents at 240V, and would cause problems to most mechanical relays. So instead of a mechanical relay, a Crydon TD2425 solid state relay was used to switch mains current. This relay is rated at 25 amps continuous and has a maximum surge current rating of 250 amps. While the welder can draw more than 25 amps, it does not do so continuously - so this relay should be fine. Similar relays are available on eBay for under $20.

Given that welding occurs for such short periods, I wondered if a heatsink would even be necessary for the solid state relay, but some load-testing of the bare relay showed that it was getting warm. To address this, I decided to mount the relay on a thick aluminium plate that would act as a heat-sink.

The solid state module is operated by a low voltage, low current source – 12V in this case.

*The finished welder with the new control panel, positioned above the welding electrodes. The panel incorporates a digital timer, digital mains voltage and current meters, a 'welding occurring' pilot light, and a manual weld over-ride switch.*

### In detail

The major electronic changes are these:

• New solid-state relay used to control mains-power switching

• Length of weld time controlled by a user-adjustable digital timer

### 2. Micro-switch

A micro-switch is the main trigger for the welding current to flow. In the original welder, a heavy-duty switch was operated by the movement of the foot pedal. This switch comprised big chunks of copper, and worked directly on the mains power supply to the transformer. To replace this, the original switch was removed and was replaced (working in much the same position) by the micro-switch, triggered by an adjustable cam plate that moves with the foot pedal. I had the micro-switch on the shelf, but similar ones cost about $4.

The micro-switch feeds power to the digital timer. Therefore, when the switch is tripped, the digital timer's output is activated, so switching on the solid state relay (and therefore the welding current) and starting the timer countdown. When the timer reaches zero, the sold state relay is turned off, ending the weld current period.

This photograph shows the welder with its rear cover removed. The new solid state relay (light arrow) controls the primary current flow to the transformer. Note the heatsinking for the relay provided by the thick aluminium plate. The dark arrow points to the micro-switch that is triggered when the foot pedal achieves full electrode clamping pressure. This switch then activates the welding timer.

### 3. Digital timer

The heart of the new control system is a digital timer.

Available very cheaply on eBay (search for "AC/DC 12V 8 Pin DPDT 0.01-9.99 Second 9s Time Delay Relay Timer Black ASY-3D"), the timer uses thumbwheel adjustable switches to set the output

time anywhere from 0.01 to 9.99 seconds. At the time of writing, it costs just $14 including postage. (Note that an 8-pin socket is required for the timer – it's purchased separately.)

When the timer is powered-up by the micro-switch, the internal output relay is energised. The digital display then counts down the timed period and when this has ended, the output relay is turned off.

Well that was the theory! Unfortunately, the timer cannot be configured in this way. Instead, the nearest function is that the relay output of the timer is on when you'd like it to be off, and off when you'd like it to be on. To achieve the required reversal in function, an automotive 12V 'normally closed' relay (about $6) was added to the circuit.

With the timer and additional relay working, the internal output relay of the timer activates the external normally closed relay, that in turn triggers the solid state relay, so feeding mains voltage to the transformer (and so a much larger secondary current to the welding electrodes). It does this for only the user-selectable timed period.

In addition, the output of the timer switches on the 'welding occurring' pilot light – this light is wired in parallel with the solid state relay control input. A 12V LED pilot light is used for this function.

In parallel with the additional relay's output is the manual over-ride welding pushbutton. With the timer set to 0, this allows the operator to determine the welding time – when the micro-switch activated by the foot pedal, welding time continues for as long as a finger is on the button.

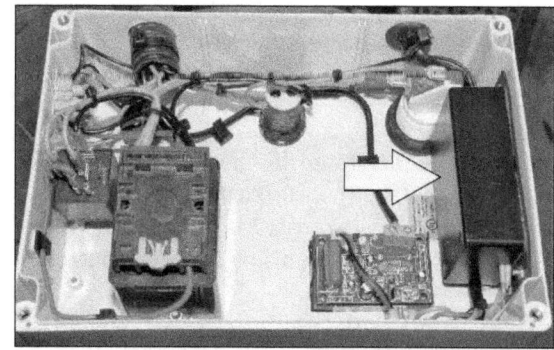

The view inside the new control panel. The arrow points to the rehoused 12V plugpack that supplies power for the electronics. At the far left is the additional relay required to reverse the timer's control logic.

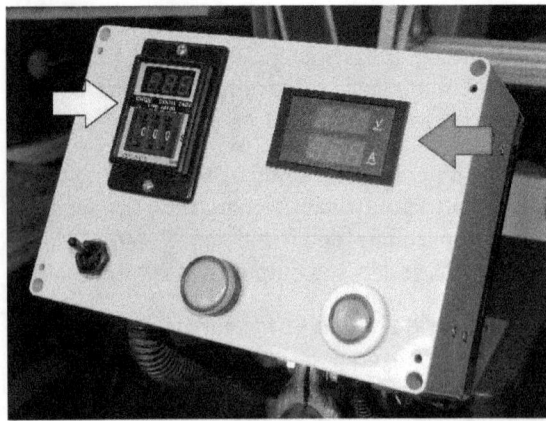

*The new control panel for the welder. The light arrow points to the adjustable digital timer that sets the duration that welding current flows. The dark arrow shows the mains voltage and current meters. Across the bottom are the on/off switch, the 'welding occurring' pilot lamps, and the manual over-ride welding pushbutton. Note that despite its 220V label, the timer is actually a 12V model.*

### 4. 12V power supply

Twelve volt power for solid state relay control, digital timer, pilot light and cooling fan is provided by the guts of a 12V regulated switch-mode plugpack, mounted within a new insulated plastic box. The plug pack was one of the salvaged ones I have on the shelf.

### 5. Mains current and voltage

The combined volt and amp meter is available on eBay – search under "AC 80-300V 100A LED Panel Digital Ammeter Voltmeter Volt Amp Meter 110 220-240V". At the time of writing it cost just $14, including postage.

The voltage is monitored by the same leads that power the meter. The current is monitored by means of a Hall Effect sensor that comprises a ring through which one of the mains power conductors is passed.

(Note that there is no electric connection between this monitoring device and the conductor.)

The output of the Hall Effect sensor is then connected to the on-board socket of the display. I used shielded cable to make this connection.

### Conclusion

The changes have revolutionised the welder. How? In some ways, not as was expected!

Firstly, the use of the electronic timer means that, especially when weld times are short, excellent results can be obtained. My test for this is to weld two mild steel roofing nails together. This welding requires precise timing – because of their low thermal mass, too long a welding period and the nails completely melt – but too short and the weld is not strong. However, with the use of the timer, consistent welds can be made each time on tricky materials such as these.

Secondly, the voltage and current displays have proved to be more useful than expected. If the contact between the materials being welded is not effective (for example, there is present residual paint, plating or corrosion), the current flow shown on the meter is much lower than normal. This immediately alerts the operator than the weld is going to be poor.

Finally, the cooling fan keeps transformer temperature noticeably lower than occurred previously.

Not long ago a project such as described here would have cost three or four times as much – making the whole welder modification a project not worth pursuing. But for an all-up project cost of around $80, the results are fantastic.

*A fan was added to circulate air around the transformer. It's funny to think that when the welder was produced over 50 years ago, fans such as these where then unknown!*

# Four-channel home sound amplifier

**A quality home sound amplifier using a mix of salvaged parts and prebuilt modules.**

In my house I installed two 15-inch woofers beneath the floor and two 8-inch two-way speakers in the walls. That's fine – but now, how to drive them?

I decided to build my own four channel amplifier to do the job. The end result cost very little (in terms of quality amplifiers, anyway) and was easy to put together. It makes extensive use of the very low cost, prebuilt eBay electronics modules.

As presented, the design requires that you use it with an input source having its own volume control (e.g. an MP3 player) or with a complete standalone pre-amp. I use a pre-amp - a Clarion EQS746 car unit run from a plug-pack. The Clarion is an excellent pre-amp available at low cost, and makes driving the four inputs from a single stereo audio source very easy.

*The amplifier has a clear top panel, internal blue LED lighting and a heatsink fan controller on the front panel.*

**Main parts**

**1. Amplifier modules**

The amplifier modules are based around prebuilt LM3886 modules. Each is capable of 68 watts into 4 ohms at a maximum distortion of 0.1 per cent. They have on-board terminal connections for line-level inputs and speaker level outputs. For best results, they require a power supply of at least plus/minus 28V at maximum load, while at the same time not exceeding plus/minus 42V at no load.

The amplifier modules are available prebuilt on eBay from about $15 each - I used four.

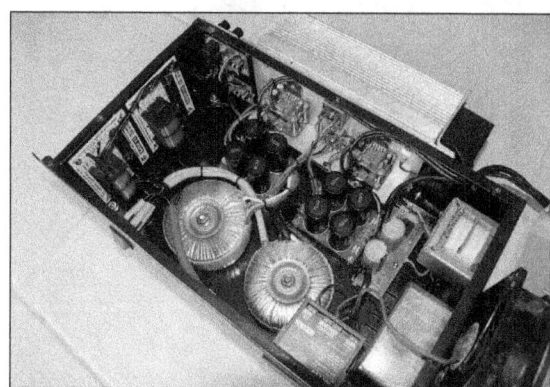

*Interior view.*

**2. Power Supply**

To feed the amplifier modules, I used prebuilt power supplies, each comprising a bridge rectifier and 4 x 10,000uF capacitors. Each power supply board can run a pair of amplifier modules.

These modules are available prebuilt on eBay from about $20 - I used two.

**3. Transformer**

Two toroidal transformers were used. Each is rated at 160VA and has a 230V primary and 25V + 25V secondaries.

When running the power supply modules (one transformer for each power supply), the measured voltage at the amplifier modules at maximum audio load was plus/minus 29V, with plus/minus 39V off-load - excellent.

The transformers can be sourced from Jaycar Electronics – they cost $55 each.

**4. Heat sink**

A huge heat sink was sourced second-hand from eBay for just $20. It was cut down a little and then bolted to a thick aluminium plate on which the four amplifier modules were mounted. The plate gave a flat surface on which to mount the ICs – the back of the heat sink had some channels cut out which otherwise made this difficult.

*Two cooling fans are used – this one is located on the end of the heatsink, and another is located in the top panel. Both are triggered by the front panel temperature controller.*

*Two speaker protection modules are used – one for each pair of channels.*

### The build

To build a working stereo (i.e. two channel) amplifier, all that you need are:

- Two amplifier modules (prebuilt)
- Power supply module (prebuilt)
- Transformer

You can see that it's easy to have an amplifier working on a breadboard very shortly after unpacking the parts!

However, in additional to running four channels rather than two, I chose to be more elaborate. I added:

- Mains switch (salvaged from other equipment)
- Box (a rack-mount triple-height unit salvaged from another amplifier)
- Front panel (eBay – new)
- IEC power socket, fuse and filter (salvaged from other equipment)

*Rear view of amplifier.*

I then further added:

- Digital temp controller displaying heat sink temp (eBay - new)
- Dual cooling fans switched by temp controller (the fans salvaged from other equipment)
- Four-channel speaker protection (eBay – two prebuilt boards)
- Internally mounted crossovers for the low frequency speakers
- Blue LED 12V lighting bar! (Jaycar Electronics)

To run the above accessories, I chose to add another power supply comprising a transformer and rectifier/filter board salvaged from other equipment – that kept the power supply for the accessories quite separate to the two powering the audio.

I chose not to use input sockets, instead using four 30cm long flying leads terminated in RCA plugs.

I also mounted new rectifiers for the power supplies on the main heatsink, rather than leaving the original ones on the boards.

## Results

Driving my underfloor 15-inch bass speakers and 8-inch in-wall two ways, the amp develops plenty of clean power. With these relatively efficient speakers, I'd never want it any louder, and distortion at full noise isn't audible to my ears.

*Added new rectifiers mounted on the heatsink.*

*The rear view of the temperature controller.*

The amplifier can be driven at maximum listening volume for hour after hour – the design copes with this just fine. Even in 30 degree Celsius ambient temps, the cooling fans cycle on and off only about every 5 minutes, the heat sink temp never exceeding 43 degrees C. At lower listening levels, or in cooler temperatures, the fans do not operate at all.

To be completely honest with you, I am amazed at how good it sounds.

*Speaker outputs.*

*This 12V transformer and associated rectifiers and capacitors runs the fans and temperature controller.*

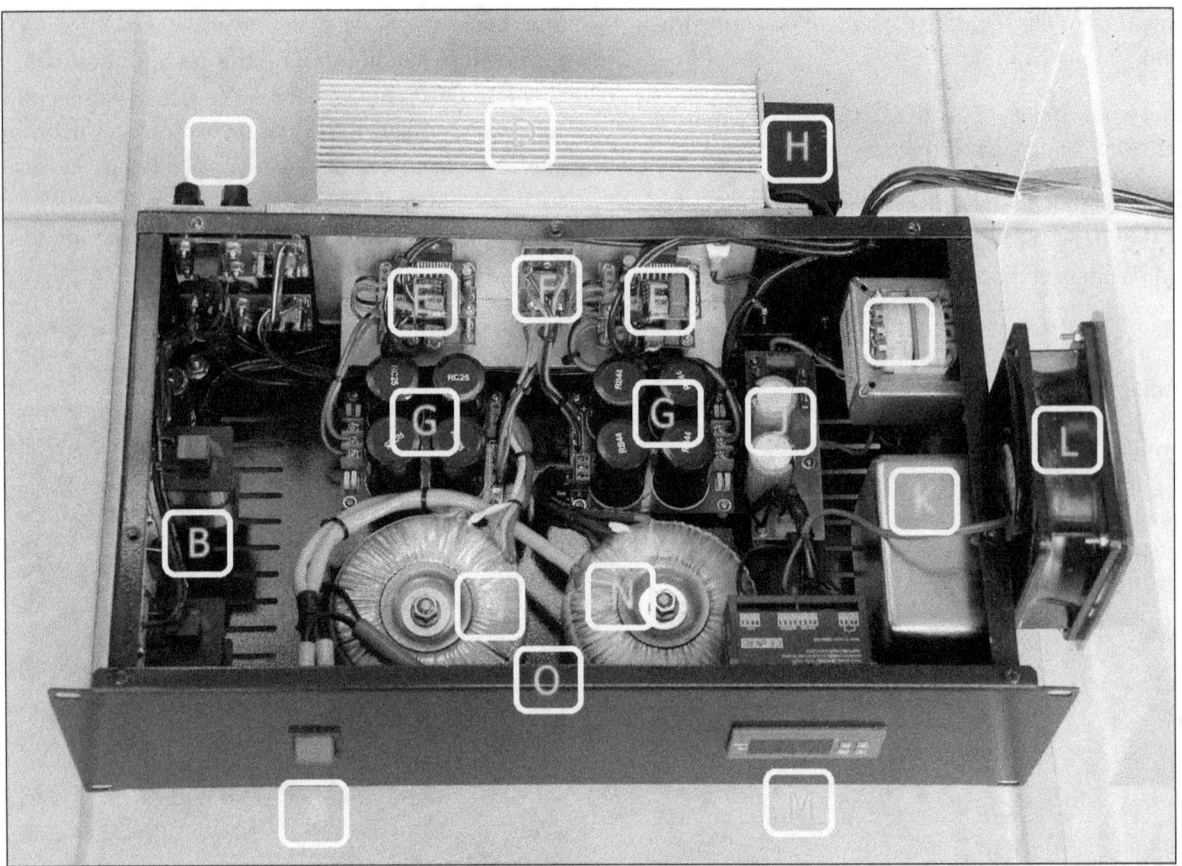

A – Power switch with inbuilt neon indicator. Salvaged from other equipment.

B – Crossovers for the underfloor bass speakers. There was room to put them in the amplifier box, so why not?

C – Speaker output terminals. These mount directly to the speaker protection modules – available prebuilt on eBay.

D – Huge heat sink, bolted to a thick aluminium plate. Heat sink sourced second-hand online.

E – Four LM3886 prebuilt modules, each developing 68 watts/channel at low distortion. Available very cheaply on eBay.

F – Power supply rectifiers. These were added as the original rectifiers, mounted on the prebuilt boards, were rather small and were not easily connected to heat sinks.

G – Power supply boards, one per pair of amplifier modules. These boards are available very cheaply on eBay.

H – Small cooling fan, mounted on the end of the heat sink. This fan and its enclosure were salvaged from other equipment.

I – 12V transformer running the speaker protection boards, temperature monitoring, cooling fans and LED lighting.

J – Rectifier and filter board for 12V transformer. This and the 12V transformer were savaged from other equipment.

K – Large mains filter, salvaged from other equipment.

L – Second cooling fan, mounted in the clear plastic lid of the enclosure. Both fans are 24V designs run at 12V to slow and quieten them.

M – LED temperature display and programmable controller. This monitors heat sink temperature and operates the fans as required. Controller and remote probe available very cheaply on eBay.

N – Two 160VA 25V + 25 V transformers, one per power supply module. One transformer was bought new (Jaycar Electronics) and the other salvaged from an old self-built amplifier.

O – Blue LED light bar mounted under front lip. Just a bit of fun!

# Four-channel car sound amplifier

**Here is a four channel, car sound amplifier with a maximum output of 68 watts per channel. That's more than enough, even with relatively inefficient car speakers, to give you plenty of volume and punchy bass. Even better, if you're happy to do some of your own metalwork and you already have some hardware like fasteners and spacers, the cost is very low. Quality? Far better than the vast majority of similar power car amplifiers!**

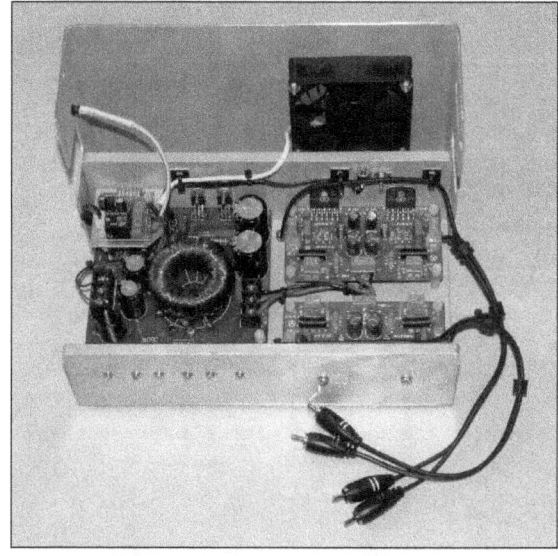

*The completed four-channel, 12V amplifier. The power supply module is on the left, and the two 2-channel amplifier modules on the right. When the lid is placed on top, it locates the fan above the amplifier modules. (A similar sized vent hole to the fan is located under the power supply module.)*

### Starting points

The heart of the amplifier comprises four LM3886 ICs. This audio amplifier IC has been around for a while – it's an oldie but a goody.

Each is capable of 68 watts into 4 ohms at a maximum distortion of 0.1 per cent.

However, rather than start with the bare ICs, we use two prebuilt, two-channel modules available on eBay. Note that the selected modules require a plus/minus 28V DC supply, rather than the AC transformer supply that most of these modules are configured for. Therefore, when sourcing these modules, ensure they look exactly as pictured.

You'll pay only about $20 for each of the two required modules – to find them, search under "Assembled LM3886TF Dual channel Stereo Audio Amplifier Board 68W+68W 4Ω 50W*2 8Ω".

Next up, you'll need a power supply capable of driving these modules.

Previously, developing such a supply would have been expensive and time consuming – but now one is available off the shelf. It's called "1PC Switching boost Power Supply board 350W DC12V to Dual ±20-32V for auto" and costs only about $36. (All prices include delivery.)

While the output of the power supply is plus/minus 32V as it arrives (the on-board pot allows adjustment), the LM3886 is happy with up to plus/minus 42V, so that's fine.

I also chose to use a fan for heatsink cooling and triggered it via another eBay module. This module is called "20-90°C DC 12V Thermostat Digital Temperature Control Switch Temp Controller New" and costs a measly $5.

### Wiring

The electronics aspect of building the amp is dead-easy.

The power supply board Input GND, K and 12V terminals are connected as indicated – GND to chassis ground, and the 12V terminal directly to the positive of the car battery. Use a high-current fuse in this battery supply – e.g. 20 amps. The K terminal requires 12V to switch on the power supply – normally it's connected to the 'power aerial' output of the head unit. (Or if you don't have this, you could connect it to any ignition-switched 12V supply.)

The power supply board's output VCC+, GND and VCC- terminals are, respectively, connected to the (+), GND and (–) terminals of the two amplifier modules.

The line level inputs from the head unit are connected to the IN amplifier module terminal blocks (observe correct polarity), and the speakers connected to the OUT terminal blocks (again observe correct polarity).

And that's it for wiring!

*This prebuilt stereo amplifier module is available on eBay for just $20, with two of these modules needed for this amplifier. The module requires at least plus/minus 28V DC to run and, in use, plenty of heatsinking is also required.*

### Building

You need to provide plenty of heatsinking capacity – either by the use of substantial heatsinks, or by using smaller heatsinks but adding a fan. In actual use, the majority of heat is generated by the four LM3886 modules – despite appearances, the power supply module heatsinking requirements are more modest.

I wanted a compact box, so made one from aluminium sheet specifically to suit the required dimensions. The overall dimensions of the box were about 250 x 140 x 75mm. The heatsinks were formed by using 8mm thick aluminium plate for two walls of the box. A salvaged 12V fan was placed in the top panel (and there is a matching size hole in the bottom panel) and the fan is triggered by the temperature sensing module. The fan is set to turn on at 40 degrees C.

*This power supply module generates plus/minus 32V DC from 12V DC and can be directly connected to the amplifier modules.*

The power supply module comes with the required insulating washers and collars for mounting the transistors to the heatsink, while the amplifier modules uses plastic encapsulated ICs and so do not require any extra insulation.

To mount the boards, you'll need to provide the insulated standoffs and screws, washers and nuts.

Rather than place connectors for the inputs and speakers on the box, I chose to directly wire these connections to the boards. These leads were run through rubber grommets that slide up appropriate channels when the lid is screwed into place.

Obviously the type of housing you place the components in is up to you – you could even use a discarded car sound amplifier enclosure that incorporates its own heatsink. But remember, whatever approach you take, you'll need either quite substantial heatsinking or need to add a fan.

### Results

The only financial outlay I had was for the eBay amplifier and power supply modules – so, $76. And for that money, this is an unbeatable amplifier. The sound is excellent - better than commercial car

*This tiny module triggers the cooling fan on the basis of the temperature selected with the DIP switches. In this case, the temperature for fan activation was set at 40 degrees C. The remote temp probe is located between two of the LM3886 amplifier ICs on one of the heatsinks.*

sound amplifiers costing two or three times much, and so much better than the typical four channel amplifier built into a head unit that it's not funny!

**Note:** this amplifier is also appropriate for anywhere you need quality four-channel sound and you're limited to a 12V supply. That includes, boats, caravans and houses working off a low voltage solar system.

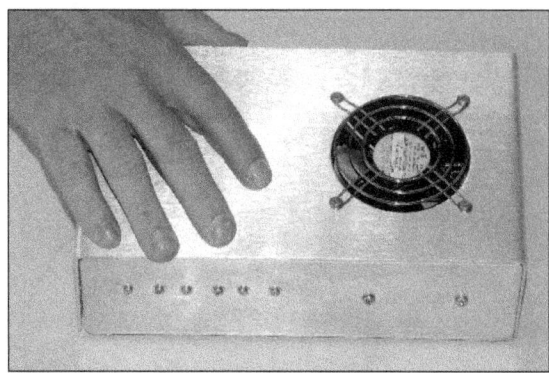

*One of the requirements was that the 270-watt amplifier be reasonably compact and light. The final item has a mass of 1.75kg and dimensions of 250 x 140 x 75mm.*

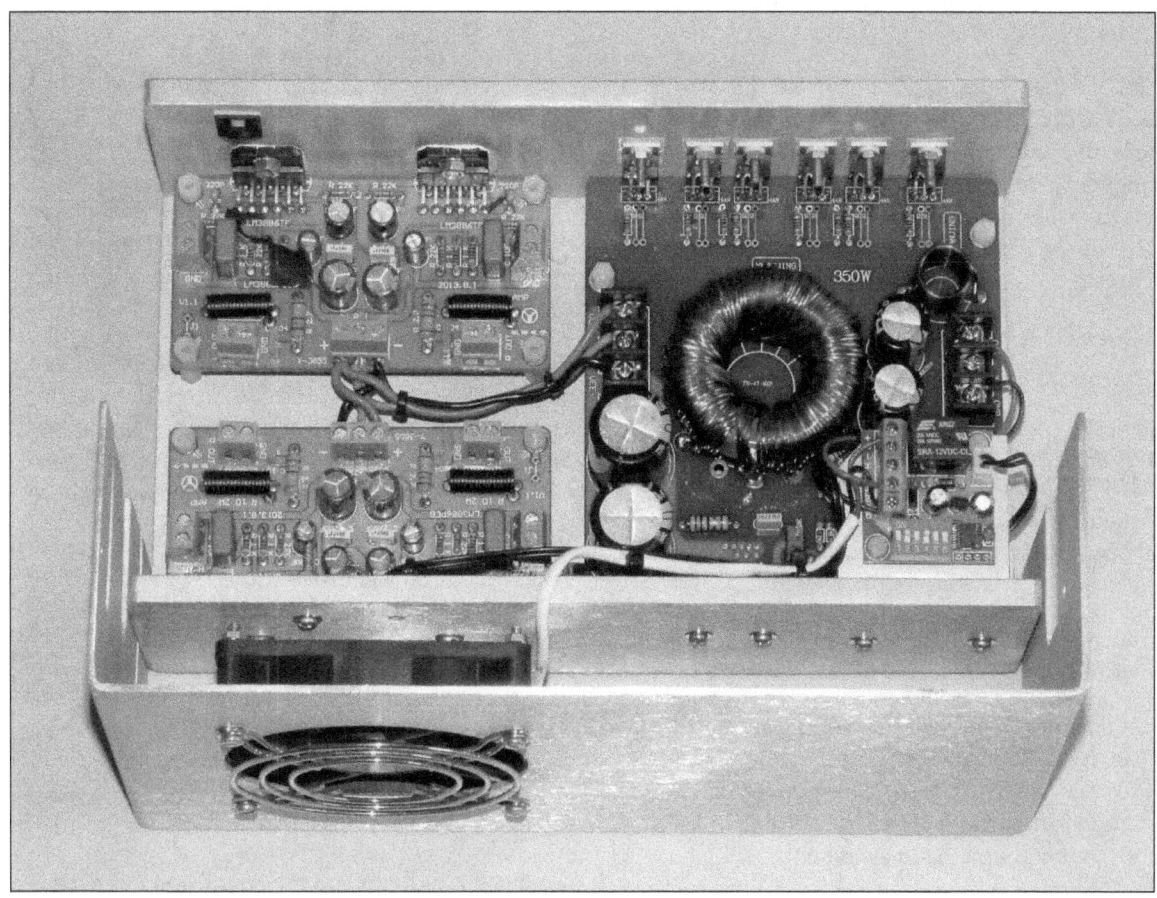

*The inside view, before speaker output leads and head unit input leads were added.*

# Variable frequency module

## Great for equipment alarms and security indicators

*This tiny variable frequency module can drive real loads like hi-intensity LEDs and piezo buzzers, and has variable frequency and duty cycle controls. It will work from 3 – 12V.*

Remember when you wanted to have a LED flasher, and built a circuit from scratch with a handful of components? And then when you wanted to change the frequency or duty cycle of the circuit, you needed to make component changes? And (for me at least!), it always took ages to get the result that was wanted.

Well, now all that has changed.

For less than the price you would have once paid for the components of that simple flasher circuit, you can now get a pre-built module that has adjustable frequency, adjustable duty cycle – and can directly drive loads like high intensity LEDs and piezo buzzers. And that price includes delivery to your letter box.

How much then? This module costs less than $4, including postage!

So what do you get for your money? The very small PCB, just 35 x 37mm, doesn't feature the IC you might expect - a 555. Instead, it uses an LM358 op amp. This in turn drives a S9012 transistor good for 0.5 amps (but more on this in a moment).

The output frequency range is adjustable from one output per 15 seconds right through to a measured

2.2 KHz. This adjustability is achieved by the use of a multi-turn pot, and links that are inserted or removed, depending on the frequency range required. The more links you place, the higher the output frequency.

The other pot adjusts duty cycle – the proportion of time the output signal is on. This pot is adjustable to give a range from 0 – 100 per cent.

With the board orientated so that the pots are on the left, the frequency pot is the lower one. Turn it clockwise to increase frequency. The upper pot is the duty cycle control; turn it clockwise to reduce duty cycle.

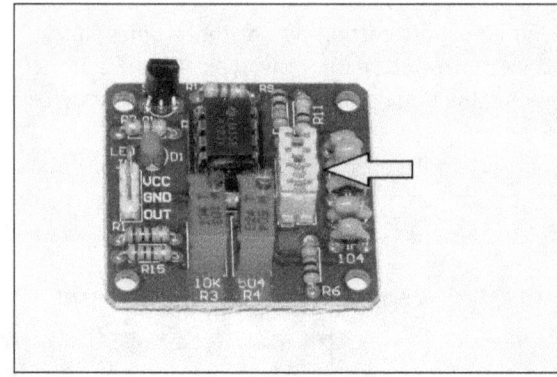

*The frequency range is selected by adding or removing these links. The range is selectable from one output per 15 seconds, right through to 2.2 KHz. Fine-tuning is then done with a pot.*

Now, if you're expecting a great-looking square wave at all frequencies and duty cycles, think again! Put a scope on it and the output can get ugly at low duty cycles, high frequencies – and pretty well everything in between. But it's not meant to be a precision frequency generator, after all.

What it does do well is to flash LEDs, pulse buzzers and perform similar functions.

For example, pulse a high intensity LED and piezo buzzer at 25Hz and no-one would overlook such a warning alarm! You can also set the duty cycle to be very short (e.g. 20 per cent) and the frequency very low (e.g. one output per 2 seconds) and in so doing, gain a very low current flasher.

A red LED is mounted on the board, and this flashes to shows the output behaviour that has been selected. If absolute minimum current draw is needed, this LED could be removed. Even with it present, with no output being driven, average current draw was about 6mA with a 9V battery powering the module.

*This is the output transistor. It is suggested that it is good for 0.5 amps but if you are driving heavy duty loads, keep a close eye on its temperature rise.*

Interestingly, when powering the high intensity LED and piezo buzzer, the low duty cycle being used meant that current draw only went up to about 12mA – a very low current demand for such an attention seeker.

For high input impedance loads, the module could also be used as a signal injector at (say) 2kHz.

I tried the module flashing a 12V, 5W LED and the output transistor started getting rather warm – and that was with a duty cycle of only about 20 per cent. So take care if you are driving larger loads.

While at least one eBay ad suggests that the module could drive relays, I could see no diode protection on the board to guard against spikes coming back from the relay coil, so I'd be a bit wary of doing this without adding an external diode.

The board worked down to just 3.0V and the specs suggest that the upper range is 12.0V. The output amplitude is similar to the power voltage, so when driving LEDs, keep in mind the necessity for a suitable dropping resistor.

In my testing, I used a 12V piezo buzzer and LED pre-wired with a resistor for 12V. This was a simple and effective combination.

# USB boost charger

*This tiny module is able to charge your phone or other USB-charged device while being powered from voltages well below the 5V USB nominal output. Search under "USB Solar Boost DC-DC 3V-5V Adjustable Power Supply Voltage Converter Module NEW".*

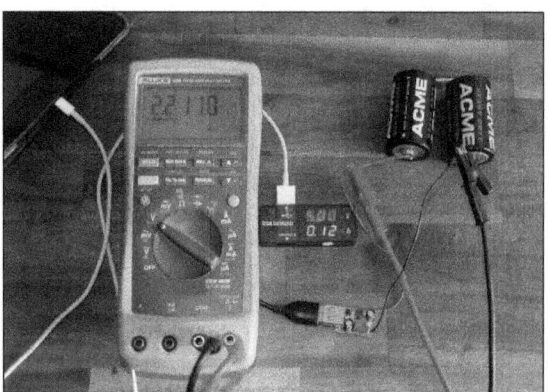

*Here's the charging module in action. Being powered by two near-flat D cells (series voltage of the pair is just 2.2V, as shown on the multimeter), the module is charging an iPad Air at 120mA and 5.0V. The tiny module is nearly lost from sight in this pic – it's at the middle, bottom of the photo.*

# Voltage booster

## Increase the output of pumps, lights and fans!

Here is an incredibly cheap voltage booster (just $6 including delivery) that has an adjustable voltage output from 12 – 35V and a decent power rating. It's rated at 150 watts (with added fan cooling), 100 watts without fan cooling (but maybe with bigger heatsinks), and we'd say on the basis of our testing, it'll be very happy at 50 – 60 watts continuous... just as it comes out of the packet.

To find it, do an eBay search under "DC-DC 10-32V To 12-35V 150W Power Supply Boost Adjust Module Mobile Laptop Car".

*The module comes as a built circuit board. It's about 65 x 50 x 30mm (L x W x H), has a heatsink along each long side, a four-terminal connection strip at one end and a multi-turn pot at the other end. It's well made – in fact, a real quality item with clear connection markings (in English), good PCB design and four tapped metal spacers (on which it sits).*

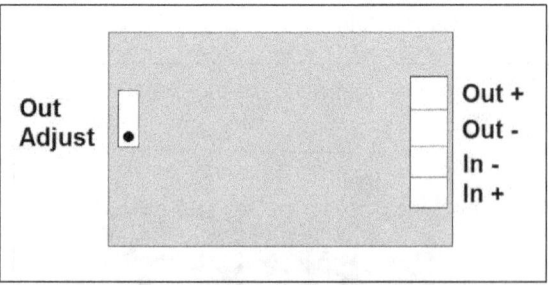

*Connections are very easy – 'IN' positive and negative, and 'OUT' positive and negative. (Don't get these connections around the wrong way, and make sure you don't short-circuit the output.) Before powering-up, turn the pot many turns anti-clockwise to reduce the gain; turn it clockwise to slowly bring up the output voltage. The output voltage is easiest measured with a multimeter. Note that this diagram is the view from above.*

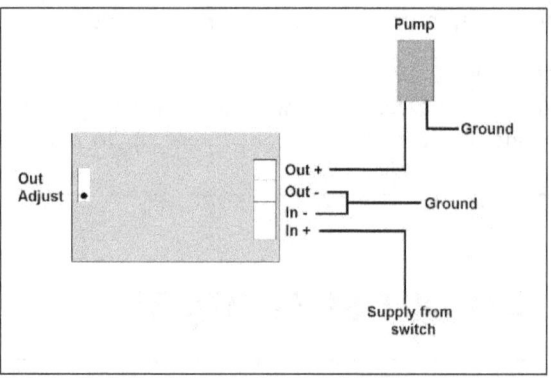

*Here's an example wiring installation running a pump. When the wiring is complete, the board should be mounted in a ventilated box.*

### Efficiency?

So how efficient is the module?

The manufacturer claims an efficiency of 94 per cent when running with an input voltage of 19V, a 2.5 amp current draw, and an output of 16V. That's not normally how you'd use it, though.

In our testing, with an input voltage of 12.0V, an output of 14.7V and a current draw of 1.25A, the efficiency was 91 per cent. In other words, the power draw was 15.0W and the output power was 13.7W – an internal loss of only 1.3W. That's pretty good.

# Variable PWM power module

### Control the brightness of lights and the speed of fans.

Here is an absolute beauty – a tiny, efficient and cheap variable pulse width power module. Hook it up to a DC fan, and you can control the fan speed via the supplied knob. Power filament light bulbs and you can control their brightness, steplessly and without flicker.

And for under $4, delivered to your letterbox, nothing beats this module for value.

*This tiny module is just 51 x 33 x 16mm but can control currents of up to 3 amps. It's perfect for controlling the speed of fans and other small motors.*

In fact, this module is so cheap that suddenly uses become viable that would otherwise require less efficient methods. The cooling fan in your amplifier a bit loud and strong? Once upon a time you'd have placed a dropping resistor in series, wasting power and generating heat. Now you can use this module instead – and via the knob you can set the fan speed precisely to give the desired flow. Want about the cheapest model train speed controller you'll ever find? Just use a discarded 12V plugpack and one of these modules – perhaps feeding the track in series with a 12V light bulb to give short-circuit protection.

So is the module cheap and nasty? Surprisingly, I found it to live up to its specs – in fact, to more than live up to them!

The module is rated by the maker at 3 amps, quite a high current considering the board's diminutive dimensions. But in testing I found it's quite happy to

run at this maximum current, with the supplied heat-sink rising only about 10 degrees C above ambient. In fact, I short-termed increased current to 4.5 amps and nothing turned to smoke...

*Connections are via a 4-way terminal block positioned at one end. The pictured heat sink is easily unscrewed and replaced with a larger one, however in our full-load testing the heat-sink rose in temperature only about 10 degrees C.*

The measured pulse width modulation frequency is 25kHz. That is, the output signal is turned on and off 25,000 times per second. This is high enough in frequency that motor windings cannot be heard 'singing', and gives good control of small DC motors. Input voltage range is 6 – 28V.

The board is about 51 x 33 x 16mm (L x W x H), with the knob protruding about 19mm. Connections are via a 4-way terminal strip, with the correction connections written on the bottom of the board in English. Four mounting holes are provided.

At this price and with this capability, this might be a situation where it's worthwhile buying half a dozen of the modules and putting them aside for use as needed. Find it by searching under "1203B DC Pulse Width Modulator PWM DC Motor Speed Controller Regulator Switch".

# Five must-haves

**Here are five eBay essentials to stock up on.**

### Battery monitoring

How often would a red/green indicator of battery voltage be useful? Lots of times, yes? Think of anything that has a battery but doesn't display battery voltage. Toys, tools, torches, radios – the list goes on.

Well, now there's a cheap pre-built answer – green for OK, red for replace (or charge) the battery.

The tiny PCB is just 30 x 17 x 10mm and uses a single red/green LED. A multi-turn pot allows you to adjust the changeover voltage to be set from 6.5 – 30V. (However, at 20V and above the module grew warm – probably better to monitor batteries in the lower voltage range.) The module has a hysteresis of about 0.2V.

Above the set point it will show green, at the set point both the green and red parts of the LED are lit, and below the set point the LED shows red.

At 12V it draws about 10mA, so for best battery economy, operate it with a momentary pushbutton – just press the button whenever you want to check battery level.

Search on eBay under 'Battery low voltage warning module Adjustable Lithium NiMH NiCD ion phosphate'. The module will cost you about $5, delivered to your letter box.

### Light and sound warning

This is another product that falls into the category – "when will I need one?", not "will I ever need one?". It's a warning light that incorporates an audible alarm. Connect 12V (either polarity) and the two LEDs will flash and the high-pitched buzzer will sound.

Note that when first triggered, the light and buzzer will be activated for a longer initial period (to get your attention), and then the follow-up flashes and sound are quicker.

The killer advantage is the price – around $8 for five of these units, again with post included. Search under "5x DC 12V 22mm Red LED Power Indicator Light Signal with Buzzer".

### High quality alligator clip hook-up leads

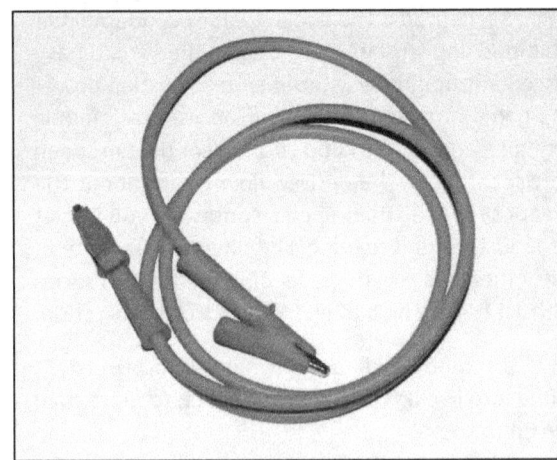

I like these so much that I've ended up buying three sets of them! So what are they? High quality, well insulated, 1-metre long leads with alligator clips at each end. And, unlike the cheap versions of these leads widely available on eBay, these are durable and will take currents of up to 5 amps. In each set you get five of the leads – green, yellow, black, red and blue.

So why are they so useful? I use them to connect chargers to batteries, to quickly and easily connect my bench power supply to circuits under test, to connect speaker drivers (again under test) to an audio oscillator – and so on. I've been using my first set continuously for more than 12 months and nothing has broken or failed.

The leads can be found by searching under "5pcs 3ft 1M high current 53mm Alligator Test Probe Clamp Clip Cable 5 Colors". They'll cost you a bit over $22. (Don't confuse them with the 50cm long leads that have very thin gauge wire and really poor alligator clips. These ones cost about $4 – they're not worth even that.)

### Adjustable voltage regulator

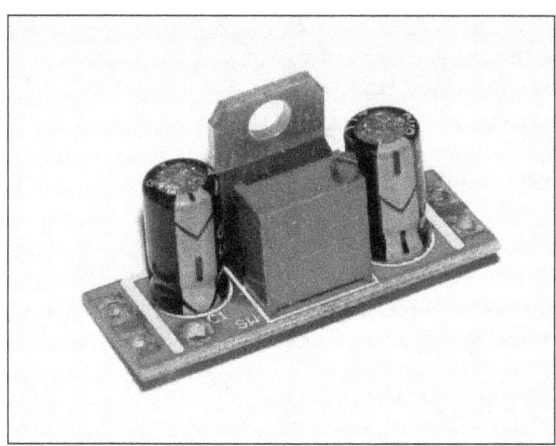

So how often do you need a low current voltage regulator? The answer to that question is: lots! So, many of us keep 3-terminal linear regulators on the shelf. They're cheap and easy – but, by the time you mount the regulator, add some capacitors (and resistors if you want to set the output level), you've spent longer than you should have. Especially when an adjustable LM317 pre-built board is available for just over $2, post included. (Or do as I did, and buy five of them to put on the shelf – just $8 total!).

The LM317 is rated at up to 1.5A output, but depending on the required voltage drop, will probably require a heatsink to achieve this. It's happy working with an output range of 1.25 to 37V, but remember that the input voltage will always

need to be about 3V higher than the required output.

The board is clearly marked with solder pads for input and output, and the multi-turn pot allows fine adjustment of the output voltage. While a heatsink is not supplied on the module I bought, it is easily attached as the regulator tab is quite accessible.

To find the module, search under 'Low Ripple Buck Linear Regulated Power Supply LM317 Module'.

Note: the output of these modules is **not** short-circuit protected!

### Potentiometer knob scales

Here are some things that I've not seen on sale recently except on eBay – they're scales that fit behind knobs on potentiometers. I thought that they looked interesting, bought some – and then I've found them in use a lot!

The scales are marked with ten-unit major increments from 0 – 100. They're 40mm in diameter and use a 10mm mounting hole. In use, you position them behind the pot's mounting nut and then rotate them until they match the sweep of the pot. The printing is good quality (protected by a plastic film that can be peeled off) and they're most effective with knobs having a maximum skirt diameter of 25mm or less.

Search under 'Rheostat Variable Transformer Potentiometer 0-100 Control Dial Face Plate 10pcs' and expect to pay just over $6 for ten of them.

# Low current power supply

## A 1-amp variable voltage power supply

You can now buy PC fan speed controllers incredibly cheaply. So? Well these controllers comprise an adjustable voltage regulator.

To get any voltage between 5 – 12V, all that you need to do is turn the knob. The voltage regulation of the one we bought was very good. When set to a 5.0V output, it maintained this with input voltages from 5.3 to 18V! And even when the load changed, the regulator continued to maintain its set output.

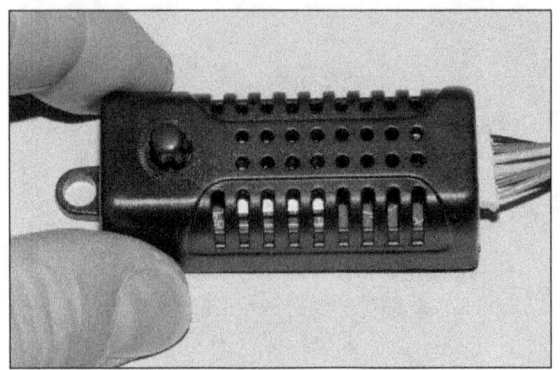

*This PC fan speed controller is an excellent low current variable power supply.*

### Making it Work

It is very easy to turn the device from a fan speed controller into a voltage regulated power supply. Two plugs are fitted as standard – these are designed to plug into the PC fan and the fan control cable.

Firstly, ensure you identify the input and output plugs correctly – the description is by the shape of the pins and not the shape of the plug!

Label the input and output. Cut off the plugs and then feed 12V to the input wires (e.g. from a plugpack) – positive to red and negative to grey. Ignore the yellow wire.

On the output side connect the red and black wires to your multimeter. Again ignore the yellow wire. Turning the knob will allow you to set the voltage. Connect the load and then check that the output voltage remains correct.

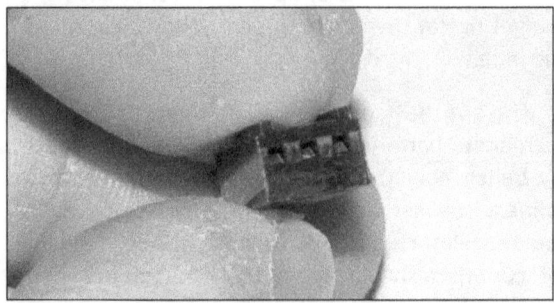

*The input plug uses female pins and on the controller we used, the input had yellow, red and grey wires.*

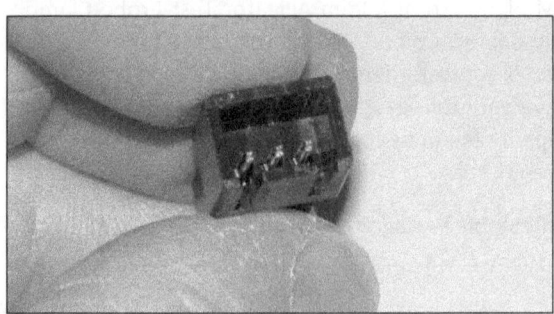

*The output plug has male pins, and used yellow, red and black wires*

### Bigger Loads

If you are driving a small digital panel meter or the like, the load current will be pretty small. However, larger loads will require a small modification to the device – fitting a larger heatsink.

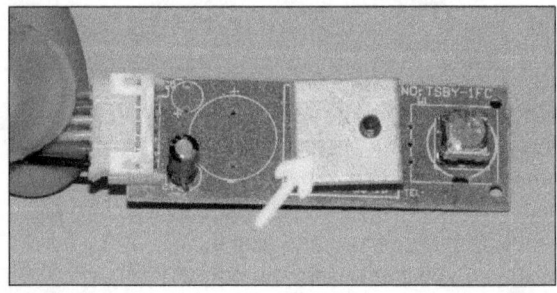

*The unit is rated to 1 amp but the internal heatsink (just a piece of square metal) is small. If you are running a current-hungry load, open-up the unit and feel the temperature of the heatsink – if it gets hot, fit a larger one. This is easy to do – just unscrew the standard heatsink and screw on a larger unit. Heatsinks are available from electronic component stores or can be salvaged from nearly any piece of discarded electronic equipment.*

# Medium current power supply

### A 2-amp variable voltage power supply for under $10!

This story is on how you can build your own bench power supply. So why would you need one?

If you do anything with DIY electronics, you need a variable voltage power supply. Turn a knob and you have a 5-volt supply, as needed by lots of electronics. Turn the knob a bit further and you have a 12V supply for testing a car radio. Turn the knob a bit further and you have 13.8V for charging a small battery – or drop it back a little and keep a car battery on float charge.

And if you have an old laptop power supply around the place, you can achieve all that for under $10 and a few hours of work.

### The parts

*The prebuilt buck converter module.*

The main building block of this design is a pre-built eBay module. Available from a number of suppliers (just search under "DC Power Supply Digital display Adjustable Voltage Buck Converter Module LM2596") the module will take any input voltage from 4-40V and turn it into a variable output from 1.25 – 37V. (If you decrease the voltage going into the module, the max output voltage also decreases.)

The module costs about $7, delivered to your letterbox. That's just stunningly cheap – especially as it includes the on-board 3-digit LED voltmeter!

The maximum peak current that the module can handle is 3 amps; it can handle 2 amps for longer periods and 1 amp continuously. If you'd like the continuous power handling figure raised, fit a heatsink to the IC.

Note that the module has short-circuit and over-temperature shutdown built in.

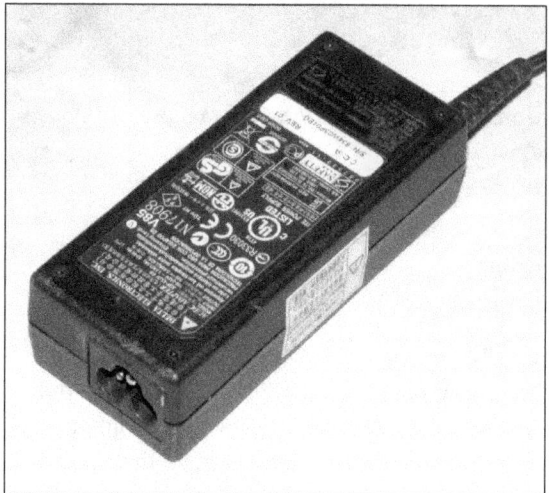

*You'll also need an old laptop power supply. One that can supply about 3 amps up to 20V is ideal.*

In addition to this module, you'll need an old ex-laptop power supply – these are readily available in currents of about 3 amps and voltages up to about 20V. Any that has anything like these specs written on it will be fine – and because they're often thrown away when a laptop is discarded, they're not hard to find.

I added an external pot to allow the voltage to be easily altered (a 20 kilo ohm unit) and a small toggle switch to allow the output to be turned on an off (often useful when you are testing a circuit and want to quickly disconnect power to make a change).

I also used a couple of output power terminals – old speaker terminals salvaged from a discarded speaker.

So that the LED voltmeter could be seen through the box, I used a red translucent food storage box purchased from a supermarket for $2.

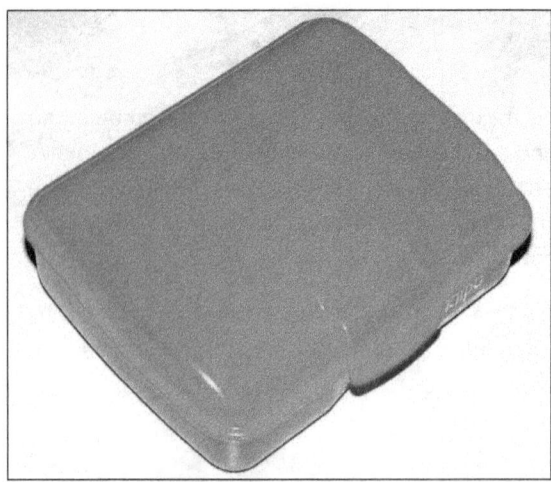

*A translucent kitchen container was used to house the power supply.*

*Remove the on-board pot, revealing its three connections.*

*Solder extension wires to these connections and run them to the new pot.*

## Building it

The first step is to ensure that the laptop power supply is working correctly. Cut off the low voltage DC plug and bare the wires. If there are more than two wires, the two thicker wires will be the power supply. Connect your multimeter to these wires, turn on mains power, and check that you have the specified voltage (e.g. 21V) on the output. Also at this stage confirm which wire is positive and which is negative.

Disconnect mains power and observing the correct polarity, connect the laptop power supply to the 'IN' terminals of the module. Plug back into mains power and check that when you rotate the on-board pot, the output voltage varies. By pressing the on-board buttons, you can turn the voltmeter on and off, and change the reading from input voltage to output voltage.

The trickiest part of the project is wiring-in the external pot. Because the printed circuit board is double-sided, it is best if you carefully use a pair of pliers to crush the on-board pot until its pieces can be removed, revealed the solder pads to which it is connected.

Carefully solder extension wires to these pads and then connect them to the external 20K pot, using the same wiring pin-outs on the new pot as were used on the old.

Reconnect mains power and check that you can vary the module output voltage by rotating the external pot.

I chose to mount a switch in the output circuit. That is, when mains power is applied to the power supply, the LED display is always illuminated. The switch just turns the output on and off.

Before assembling the box, I set the module pushbuttons to switch the LED voltmeter on, and to configure it to show output voltage.

The module is mounted using stand-offs formed by screws and multiple nuts, inserted from the front.

With the wiring completed, I used double-sided tape to stick the translucent box to the ex-laptop power supply, positioning the two 'blocks' so that the display and controls sit angled upwards.

*The completed power supply.*

In use the power supply works very well. The regulation is fairly good – the output voltage doesn't change too much when you connect a load. The supply will also cope with short-term short circuits.

However, you can't set the output voltage down to very small increments – because the original on-board pot was a 10-turn unit and we're using a one-turn pot, the fine resolution of the original control isn't retained. That said, on the prototype and using a 3.42 amp, 19V power supply, you can set the output at about 0.2 volt intervals from 1.1V to 18V.

This is a great project – cheap, very useful, rugged and (for beginners in electronics) quite safe to make.

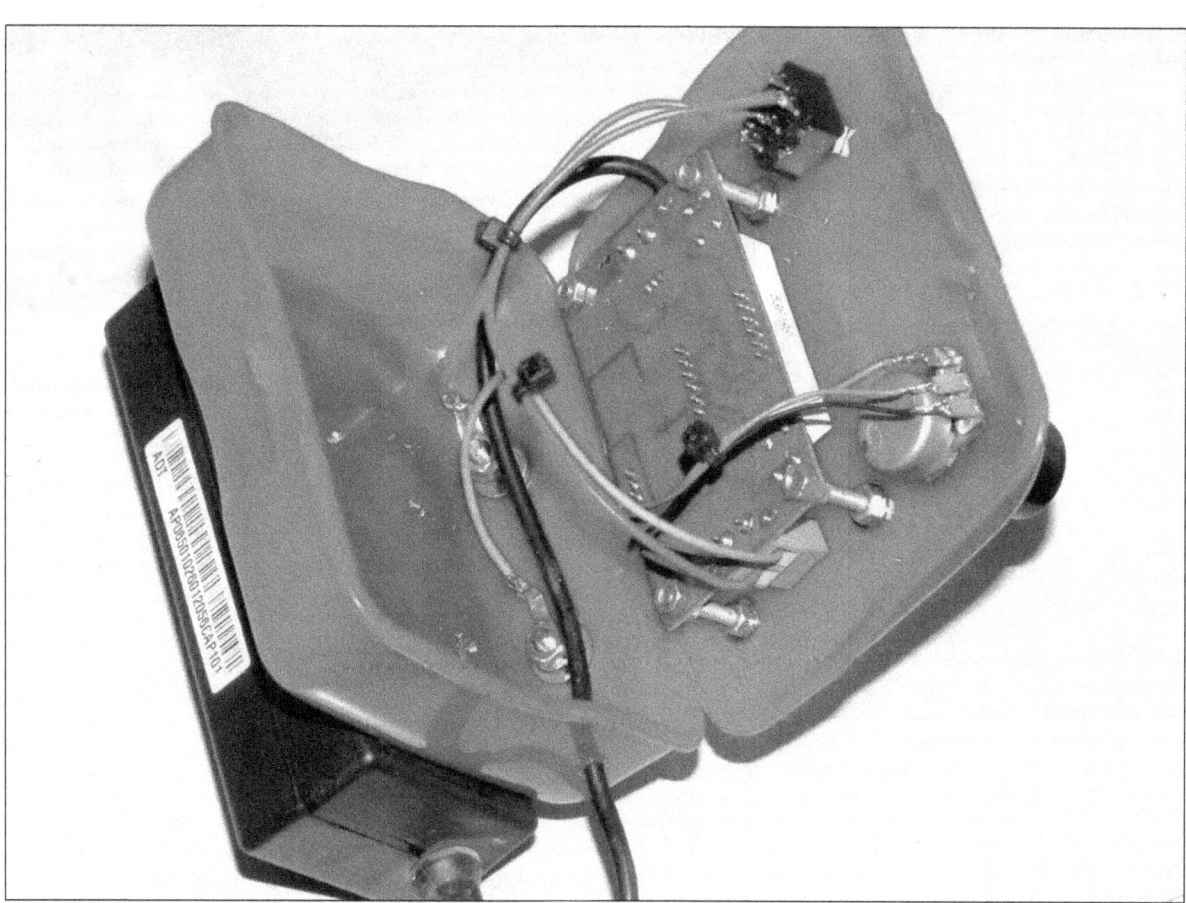

*Overall inside view.*

# High current power supply

**Here's a cheap variable power supply that can provide up to 5 amps and voltages up to 25V. Depending on how you source the parts, it can cost you under $20.**

Here's a variable voltage power supply capable of lots of current. How much then? Well, depending on how you build it, up to a short-term output of 8 amps at 12 volts, and a continuous 5 amps at around the same voltage. But is it limited to 12V output? No! You should be able to crank the output to 25V without issues.

### What you need

To make this design really cost-effective, you'll first need an old laptop or industrial power supply module.

Laptop power supplies are readily available in currents of about 3 amps and voltages up to about 20V. Any that has anything like these specs written on it will be fine – and because these power supplies are often thrown away when a laptop is discarded, they're not hard to find.

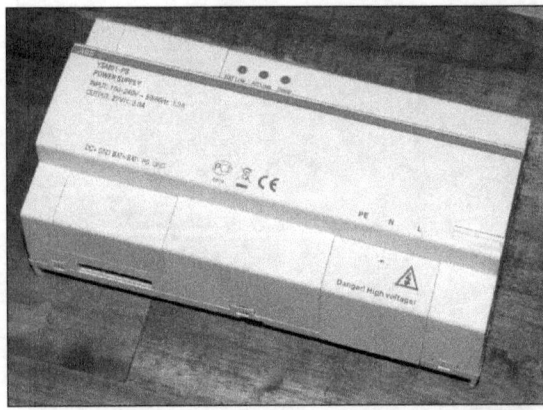

*To make this design really cost-effective, you'll first need an old laptop or industrial power supply module. I managed to pick up this brand new ABB YSM01-PS power supply module for chickenfeed ($5!) on eBay. This module can develop up to 3 amps at 27V.*

Industrial power supply modules (like the one used here) are commonly available with currents up to 3 amps at 27 volts, and – because of their solid-state design – can produce higher currents at lower voltages. For example, the one shown here developed an easy 5 amps at 12V.

Note that the max voltage you're looking for must be under 32V.

Basically, keep your eyes open for discards or cheap sale prices on these components – and the higher the spec'd max current, the better.

### Building block

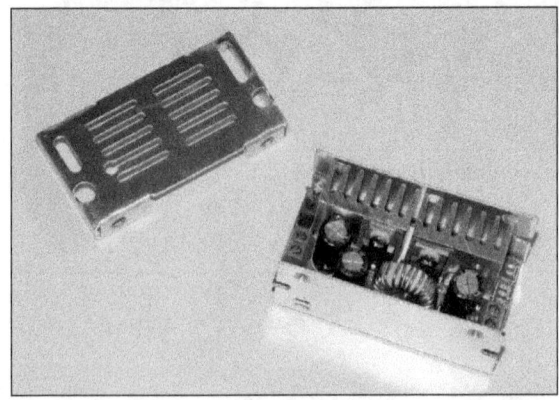

*The heart of the power supply is this eBay adjustable buck converter module.*

The main building block of this design is a pre-built eBay module. Available from a variety of sellers for around $9 (including postage!), it's called a "DC-DC Step Down 12A 200W Adjustable Converter Buck Module 4.5-30V to 0.8-32V". As its name suggests, you can feed it any DC voltage up to 32V and it will turn it into a lower voltage. If the input is 32V, the output can be anything from 0.8 to 30V. (If the input is lower, so too will be the maximum output.)

(Note that the module states that it is good for 12 amps, but more on this in a moment!)

But here's an important point. There are lots of 'buck' converters available on eBay but many of them are very fragile – they work for a few moments and then when overloaded or otherwise mistreated, they die. In fact, in research for this design, three other eBay buck converter designs were trialled but none proved to be up to the task. Ensure the module you buy looks exactly like the one pictured!

To the module I added an external 10-turn pot (a 50 kilo ohm unit) to allow the voltage to be easily and precisely altered, and a toggle switch to allow the output to be turned on an off (often useful when you are testing a circuit and want to quickly disconnect power to make a change). If you don't have these parts, they're available from electronic stores like Jaycar. I also added a cheap eBay voltmeter (a 2-wire design) and mounted the lot in a kitchen-style sealed container – a very cheap but durable box.

Because builds will vary depending on what parts are available, here's just an overview of how I did it.

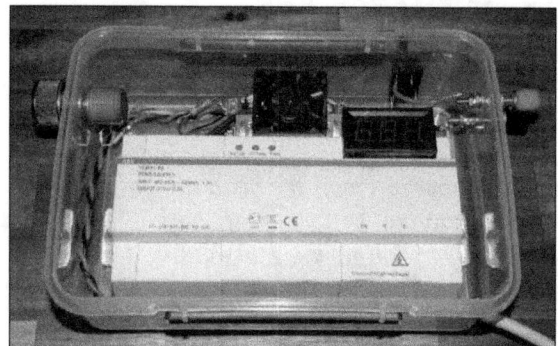

The power supply was mounted in a plastic kitchen container.

If you don't want to make any changes to the module at all, the voltage output can be altered by using a small screwdriver on the pot (arrowed), and the module turned on and off by connecting or disconnecting power appropriately.

The buck converter was taken from its metal box (the box adds nothing to the heatsinking)...

... and then, because I wanted to replace the on-board pot with an external one, the existing pot was crunched with a pair of pliers until it fell apart, revealing its connections. (Taking this approach is easier than unsoldering the pot and then soldering to the PCB pads.) The wires to the new pot were then soldered to these wire connections.

The new pot. Note that because the pot I had wasn't the right value (you need 50 kilo-ohms) I added some external resistors to correct the value – if you pick a 50 kilo-ohm pot, you won't need to do this.

A toggle switch was added and two 'binding post' style output terminals were used. Note that I chose to switch only the low voltage output – the mains power switching is done by turning the power supply on and off at the power point.

To monitor output voltage, I wired a low cost eBay voltmeter to the output. These 2-wire designs are very easy to connect, although note that they won't work down to 0 volts – about 3V is as low as they go.

Testing showed that when running continuously at currents of over 5 amps, the heatsinks on the buck converter started getting quite hot. I added a salvaged fan – but unless you want to drive high current loads continuously, you won't need a fan. When the current load is really high, I unclip the box lid to allow air to better circulate.

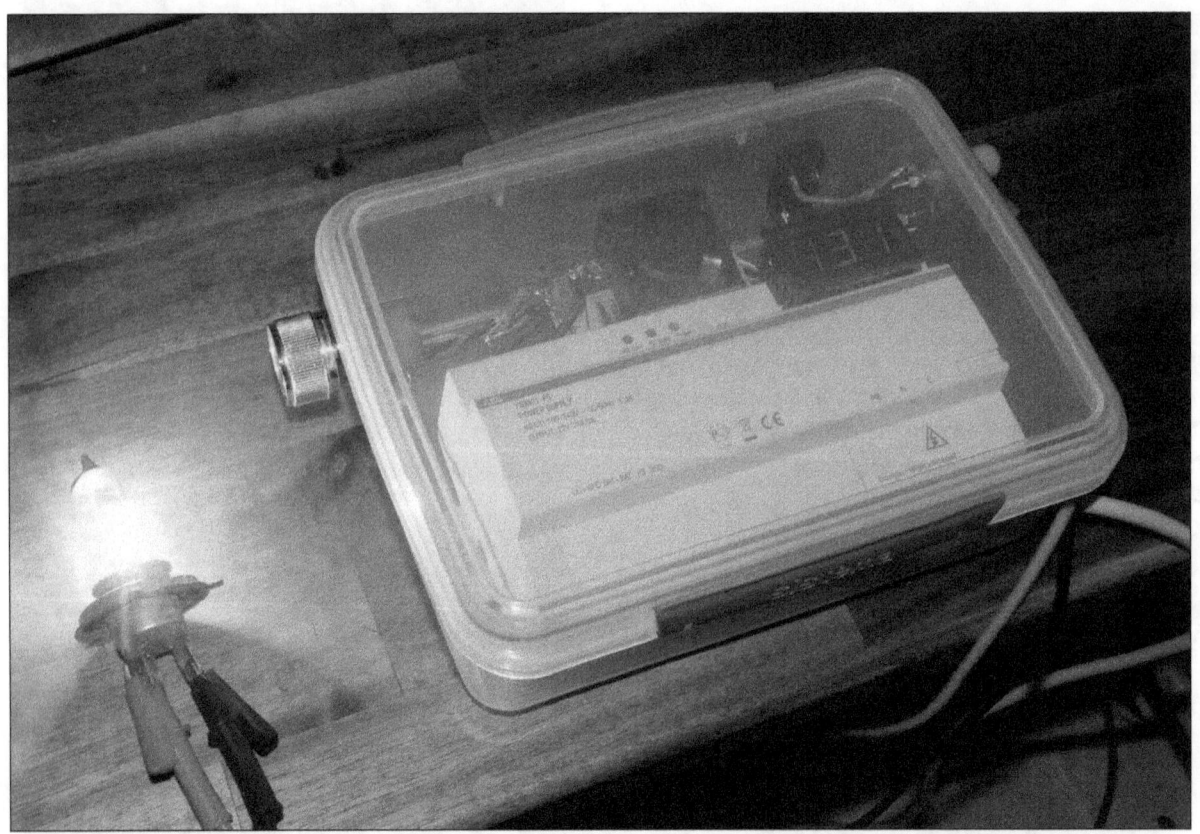

Here's the completed power supply running a 50W automotive light bulb at 13.1V. Voltage regulation is good (i.e. it doesn't vary much when you add or disconnect the load) and the module is over-load, over-temp and short-circuited protected (although you may also want to add an external fuse).

# Classic car immobiliser

### Under $12 for a remote-controlled immobiliser!

Over the last 30-odd years, the level of security built into cars has skyrocketed. Without having the right key, or a lot of time and electronic ability, thieves find stealing modern cars difficult. But that's not the case for older cars. So what if you have an older car that you want to make less stealable? And especially less stealable to an opportunistic thief? Well, for less than $12 and 30 minutes of your time, you can now add a remote control immobiliser to your older car.

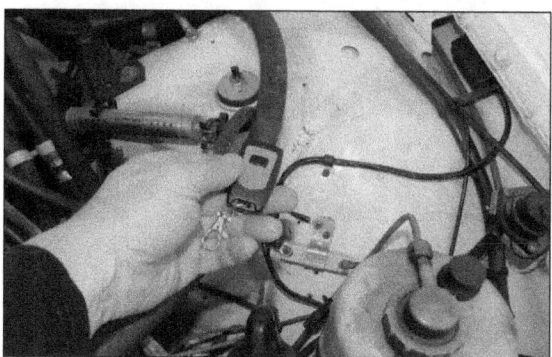

### The system

*The key remote has a single pushbutton on it, while the associated box contains the receiver and a relay output. The module plus remote is sold on eBay as the "12V Multi-function Learning Remote Control Switch New".*

*Inside the receiver box you'll find a small circuit board with a few LEDs, a configurable link and a 5-terminal strip.*

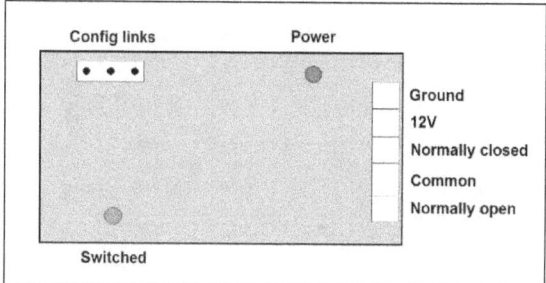

*The functions of the important bits are as shown here.*

By changing the config link position you can set the device in the following ways. (Note that this is with the board orientated as shown in this diagram and viewed from above.)

• No link – relay closes and green LED turns on only when remote button is being constantly pressed

• Link set to left - relay closes and green LED turns on when remote button is pushed once, then stays on until power to the receiver is removed

• Link set to right – relay closes and green LED turns on when remote button is pushed once, stays on until button is pushed again

Having these different configs available makes the unit very useful.

**Wiring**

To disable the car, you'll need first to decide on a wire that, if broken, will stop the car starting. A further caveat is that unless you add another relay, that wire should pass less than 10 amps when the car is running. (The remote receiver's relay is rated at 10 amps.)

The wire you choose to intercept could be to an electric fuel pump or the ignition coil.

In the case of this 1979 Mercedes, I chose to disable the ignition coil. Furthermore, I mounted the receiver module under the bonnet and powered it from the wire I was disabling – an approach that made the wiring very simple.

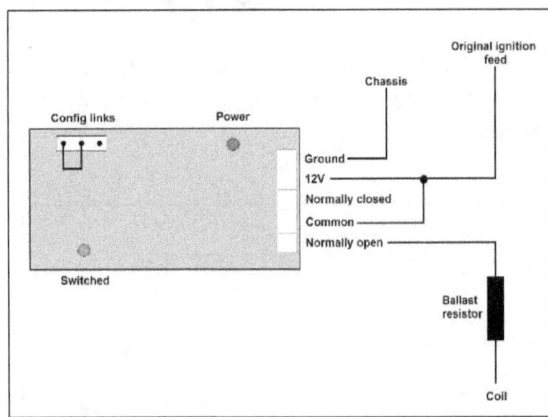

*Suggested wring diagram.*

Here are the steps:

1. Set the config link as shown – this means the relay will pull-in when the remote button is pressed and will stay latched until power to the unit is removed.

2. Find the 12V feed to the coil. Note that if the car uses a ballast resistor, you will need to work ahead of the resistor as shown here.

3. When the relay closes, the 'common' and 'normally open' connections are joined, so as shown here, when the relay is closed, power gets fed to the ballast resistor (and so then to the coil).

4. Power for the module is gained from the ignition-switched 12V source feeding the ballast.

5. You need to add a new ground wire for the module – connect this to the chassis.

Wired in this way, the steps in starting the car as follows:

1) Turn ignition key until dash warning lights are lit

2) Press remote button

3) Turn key further to start engine

Note that pressing the remote with the car switched off doesn't do anything – rather deceptive for a thief! Note also that you don't need to press the remote button to activate the immobiliser when you leave the car – as soon as you turn the ignition off, the relay automatically opens.

*Here are the coil (dark arrow) and ballast resistor (light arrow) in the Mercedes.*

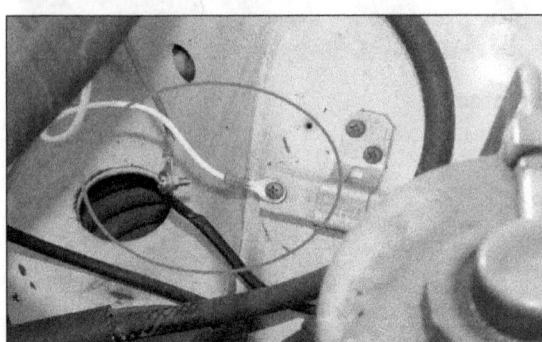

*The power supply lead to the ballast resistor was parted at the terminal and extensions put in place to connect to the immobiliser module.*

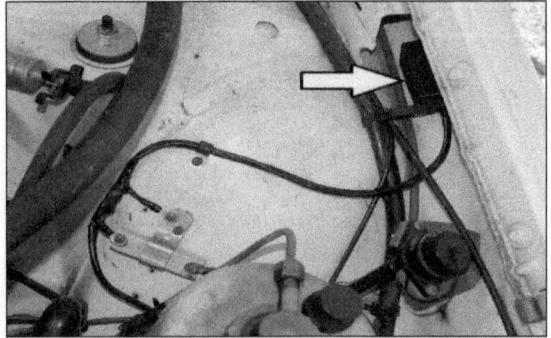

*The fitted system, with the receiver module arrowed. Seal the module's box with some neutral-cure silicone.*